T0073573

WHY
WE
DIE

Also by Venki Ramakrishnan

Gene Machine: The Race to Decipher the Secrets of the Ribosome

WHY WE DIE

THE NEW SCIENCE OF
AGING AND THE QUEST
FOR IMMORTALITY

Venki Ramakrishnan

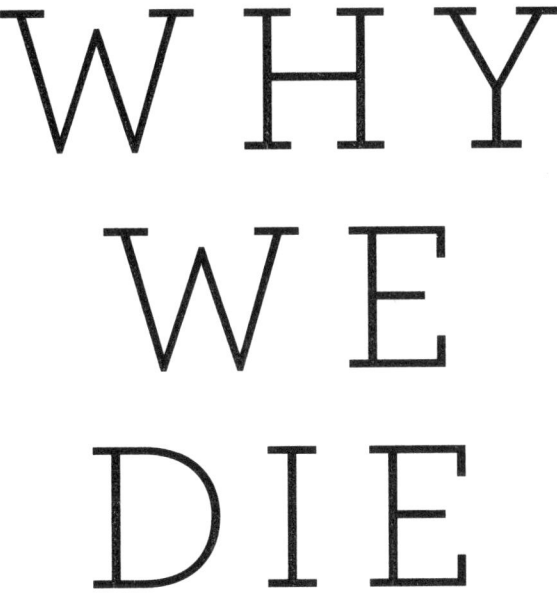

WILLIAM MORROW

An Imprint of HarperCollins*Publishers*

WHY WE DIE. Copyright © 2024 by Venki Ramakrishnan. All rights reserved. Printed in the United States of America. No part of this book may be used or reproduced in any manner whatsoever without written permission except in the case of brief quotations embodied in critical articles and reviews. For information, address HarperCollins Publishers, 195 Broadway, New York, NY 10007.

HarperCollins books may be purchased for educational, business, or sales promotional use. For information, please email the Special Markets Department at SPsales@harpercollins.com.

FIRST EDITION

Designed by Nancy Singer
Illustrations by Elfy Chiang

Library of Congress Cataloging-in-Publication Data

Names: Ramakrishnan, Venki, 1952– author.
Title: Why we die : the new science of aging and the quest for immortality / Venki Ramakrishnan.
Description: First edition. | New York, NY : William Morrow, [2024] | Includes bibliographical references and index. | Summary: "A groundbreaking exploration of the science of why and how we age and die—from Nobel Prize-winning molecular biologist Venki Ramakrishnan" —Provided by publisher.
Identifiers: LCCN 2023033370 (print) | LCCN 2023033371 (ebook) | ISBN 9780063113275 (hardcover) | ISBN 9780063113299 (ebook)
Subjects: LCSH: Death. | Death (Biology) | Death—Causes.
Classification: LCC QP87 .R34 2024 (print) | LCC QP87 (ebook) | DDC 616.07/8—dc23/eng/20231129
LC record available at https://lccn.loc.gov/2023033370
LC ebook record available at https://lccn.loc.gov/2023033371

ISBN 978-0-06-311327-5

24 25 26 27 28 LBC 5 4 3 2 1

For Vera,

my companion in aging

CONTENTS

WHY
WE
DIE

INTRODUCTION

Almost exactly one hundred years ago, an expedition led by the Englishman Howard Carter unearthed some long-buried steps in the Valley of Kings in Egypt. The steps led to a doorway with royal seals, signifying that it was the tomb of a pharaoh. The seals were intact, meaning that nobody had entered for more than three thousand years. Even Carter, a seasoned Egyptologist, was awestruck by what they found inside: the mummified young pharaoh Tutankhamun, with his magnificent gold funerary mask, kept company in the tomb for millennia by a wealth of ornate and beautiful artifacts. The tombs had been secured shut so that mere mortals could not enter—the Egyptians had gone to enormous efforts to create objects never intended to be seen by other people.

The splendor of the tomb was part of an elaborate ritual aimed at transcending death. Guarding the entrance to a room of treasures was a gold and black statue of Anubis, the jackal-headed god of the underworld, whose role is described in *The Egyptian Book of the Dead*. A scroll of the book was often placed in the pharaoh's sarcophagus. We may be tempted to think of it as a religious work, but it was more akin to a travel guide, containing instructions for navigating the treacherous underworld passage to reach a blissful afterlife. In one of the final tests, Anubis weighs the heart of the deceased against a feather. If the heart is found to be heavier, it is impure, and the person is condemned to a horrible fate. But if the examinee is pure, he would enter a beautiful land filled with eating, drinking, sex, and all the other pleasures of life.

The Egyptians were hardly alone in their beliefs of transcending death with an eternal afterlife. Although other human cultures may not have constructed such elaborate monuments as the Egyptians did for their royalty, all of them had beliefs and rituals around death.

It is fascinating to consider how we humans first became aware of our mortality. That we are aware of death at all is something of an accident, requiring the evolution of a brain that is capable of self-awareness. Very likely it needed the development of a certain level of cognition and the ability to generalize as well as the development of language to pass on that idea. Lower life forms and even complex ones such as plants, don't perceive death. It simply happens. Animals and other sentient beings may instinctively fear danger and death. They recognize when one of their own has died, and some are even known to mourn them. But there is no evidence that animals are aware of their own mortality. I do not mean being killed by an act of violence, an accident, or a preventable illness. Instead, I mean the *inevitability* of death.

At some point, we humans realized that life is like an eternal feast that we join when we are born. While we are enjoying this banquet, we notice others arriving and departing. Eventually it is our turn to leave, even though the party is still in full swing. And we dread going out alone into the cold night. The knowledge of death is so terrifying that we live most of our lives in denial of it. And when someone dies, we struggle to acknowledge that straightforwardly, and instead use euphemisms such as "passed away" or "departed," which suggest that death is not final but merely a transition to something else.

To help humans cope with their knowledge of mortality, all cultures have evolved a combination of beliefs and strategies that refuse to acknowledge the finality of death. Philosopher Stephen Cave argues that the quest for immortality has driven human civilization for centuries. He classifies our coping strategies into four plans. The first, or Plan A, is simply to try to live forever or as long as possible. If that fails, then Plan B is to be reborn physically after you die. In Plan C, even if our body decays and cannot be resurrected, our essence continues as an immortal soul. And finally, Plan D means living on through our legacy, whether that consists of works and monuments or biological offspring.

All of humanity has always incorporated Plan A into their lives, but cultures differ in the extent to which they fall back on the other plans. In India, where I grew up, Hindus and Buddhists gladly embrace Plan C, and the idea that each person has an immortal soul that lives on after death by being reincarnated in a new body, even in a completely different species. The Abrahamic religions, Judaism, Christianity, and Islam, subscribe to both Plans B and C. They believe in an immortal soul but also in the idea that we will rise *bodily* from the dead and be judged at some point in the future. Perhaps this is why traditionally these religions insisted on burial of the intact body and forbade cremation.

Some cultures, such as the ancient Egyptians, hedged their bets by incorporating all four plans into their belief systems. In grandiose tombs, they mummified the corpses of their pharaohs so that they might rise up bodily in the afterlife. But they also believed in a soul, called Ba, that represents the essence of the person and survives death. The first emperor of a unified China, Qin Shi Huang, took a similarly multipronged approach to immortality. Having escaped many attacks on his life, conquered warring states, and consolidated his power, he turned his attention to seeking the elixir of life. He sent emissaries to pursue even the faintest rumors of its existence. Facing certain execution for their failure to find it, many quite sensibly absconded and were never heard from again. In an extreme combination of Plans B and D, Qin also ordered the construction of a city-sized mausoleum for himself in Xian, employing 700,000 men in the process. The tomb contained an army of 7,000 terra-cotta warriors and horses—all meant to guard the deceased emperor until he could be reborn. Qin died at the age of forty-nine in 210 BCE. Ironically, it may have been toxic potions taken to prolong his life that ultimately cut it short.

Our ways of coping with death began to change with the arrival of the Enlightenment and modern science in the eighteenth century. The growth of rationality and skepticism means that although many of us still hang on to some forms of Plans B and C, deep down we have become less sure they are real alternatives. Our focus has shifted toward finding ways to stay alive and preserving our legacy after we die.

It is a curious facet of human psychology that even if we accept that we ourselves will be gone, we feel a strong need to be remembered. Today, instead of constructing tombs and monuments, the very rich engage in philanthropy, endowing buildings and foundations that will long outlast them. Throughout the ages, writers, artists, musicians, and scientists have sought immortality through their works. Ultimately, however, living on through our legacy is not an entirely satisfying prospect.

If you are neither a powerful monarch or billionaire, nor an Einstein, do not despair. The other way to leave a legacy and be remembered is accessible to nearly all living things, which is to live on through our offspring. The desire to procreate so that some part of us will live on is one of the strongest biological instincts to have evolved, and is so central to life that we will have much more to say about it later. But even though we love our children and grandchildren and want them to live on long after we are gone, we know that they are separate beings with their own consciousness. They are not us.

Nevertheless, most of us do not live in constant existential angst about our mortality. Rather, our brains appear to have evolved a protection mechanism by thinking of death as something that happens to other people, not ourselves. A separation of the dying reinforces the delusion. Unlike the past, when we were confronted by people dying all around us, today people often die in care homes and hospitals, isolated from the rest of the population. As a result, most of us, especially young people, go about our daily lives acting as though we are immortal. We work hard, engage in hobbies, strive after long-term goals—all useful distractions from potential worry about dying. However, no matter what tactics we employ, we cannot fully escape awareness of our mortality.

And that brings us back to Plan A. The one strategy that all sentient beings have had in common for millions of years is simply to try to stay alive for as long as possible. From a very young age, we instinctively avoid accidents, predators, enemies, and disease. Over millennia, that universal desire led us to protect ourselves from attacks by forming communities and fortifications and developing weapons and maintaining armies; but it also led to the search for potions and cures and eventually to the development of modern medicine and surgery.

For centuries, our life expectancy hardly changed. But over the last 150 years, we have doubled it, primarily because we better understood the causes of disease and its spread, and improved public health. This progress allowed us to make enormous strides in extending our average life span, largely as a result of reducing infant mortality. But extending maximum life span—the longest we can expect to live even in the best of circumstances—is a much tougher problem. Is our life span fixed, or could we slow down or even abolish aging as we learn more about our own biology?

Today the revolution in biology that began with the discovery of genes more than a hundred years ago has led us to a crossroads. For the first time, recent research on the fundamental causes of aging is raising the prospect not merely of improving our health in old age but also of extending human life span.

Demographics is driving a huge effort to identify the causes of aging and to find ways to ameliorate its effects. Much of the world is faced with a growing elderly population, and keeping them healthy for as long as possible has become an urgent social imperative. The result is that after a long period in which it was a scientific backwater, aging research—or gerontology—has taken off.

In the last ten years alone, more than 300,000 scientific articles on aging have been published. More than 700 start-up companies have invested a combined many tens of billions of dollars to tackle aging—and this is not counting large, established pharmaceutical companies that have programs of their own.

This enormous effort raises a number of questions. Could we eventually cheat disease and death and live for a very long time, possibly many times our current life span? Certainly some scientists make that claim. And California billionaires, who love their lifestyles and don't want the party to end, are only too willing to fund them.

The immortality merchants of today—the researchers who propose trying to extend life indefinitely and the billionaires who fund them—are really a modern take on the prophets of old, promising a long life largely free of the fear of encroaching old age and death. Who would have this life? The tiny fraction of the population who could afford it? What would be the ethics of treating or modifying humans to achieve this? And if it becomes widely available, what sort of society would we have? Would we be sleepwalking into a future without considering the potential social, economic, and political consequences of humans living well beyond our current life spans? Given recent advances and the enormous amount of money pouring into aging research, we must ask where this research is leading us, as well as what it suggests about the limits of human beings.

The coronavirus pandemic that hit the world in late 2019 is a stark reminder that nature does not care about our plans. Life on Earth is governed by evolution, and we are yet again reminded that viruses have been here long before humans, are highly adaptable, and will be here long after we are gone. Is it arrogant to think that we can cheat death using science and technology? If it is, what should our goals be instead?

I have spent most of my long career studying the problem of how proteins are made in the cells that make up our body. The problem is so central that it impinges on virtually every aspect of biology, and over the last few decades, we have discovered that much of aging has to do with how our body regulates the production and destruction of proteins. But when I started my career, I had no idea that anything I did would be connected with the problem of why we age and die.

Although fascinated by the explosion in aging research that has led to some very real breakthroughs in our understanding, I have also watched with growing alarm the enormous amount of hype associated with it, which has led to widespread marketing of dubious remedies that have a highly tenuous connection with the actual science. Yet they continue to flourish because they capitalize on our very natural fear of growing old and disabled and eventually dying.

That natural fear is also the reason that growing old and facing death is the subject of innumerable books. They fall into a few categories. There are books that provide practical advice on how to age healthily; some are sensible, while others border on snake oil. Others are about how to face our mortality and accept our end gracefully. These serve both a philosophical and moral purpose. Then there are books that delve into the biology of aging. These too fall into a couple of categories. They are written either by journalists or by scientists who have considerable personal stake in the form of their own start-up anti-aging companies. This book is not any of these.

Considering how rapidly the field is advancing, the enormous amount of both public and private money invested in it, and the resulting hype, I thought it was an appropriate moment for someone like me, who works in molecular biology but has no real skin in the game, to take a hard, objective look at our current understanding of aging and death. Because I know many of the leading figures in this area personally, I have been able to have many frank conversations to gain an honest and deeper understanding of how they see aging research in its many aspects. I have deliberately refrained from talking to those scientists who have made their positions clear in their own books, especially when they are also tied closely to commercial ventures on aging, but I have discussed their highly publicized views.

Given the pace of discovery, any book that focuses just on the most recent aging research would be out of date even before it was published. Moreover, the most recent discoveries in any area of science often do not hold up to scrutiny and have to be revised or discarded. Accordingly, I have tried to concentrate on some of the essential principles behind the most promising approaches to understanding and tackling aging. These principles should not only stand the test of time, but also help readers realize how we got to our present state of knowledge. I also give a historical background to some of the basic research that led to our current understanding. It is both fascinating and important to realize how much of what we know began with scientists studying some completely different fundamental problem in biology.

I said I have no skin in the game, but, of course, all of us do. We are all concerned about how we will face the end of life—less so when we are young and feel immortal, but more so at my age of seventy-one, when I find that I can do only with difficulty, or not at all, things I could do easily even just ten or twenty years ago. It sometimes feels that life is like being constrained to a smaller and smaller portion of a house, as doors to rooms that we would like to explore slowly close shut as we age. It is natural to ask what the prospects are that science can pry those doors open again.

Because aging is connected intimately with so many biological processes, this book is also something of a romp through a lot of modern molecular biology. It will take us on a journey through the major advances that have led to our current understanding of why we age and die. Along the way, we will explore the program of life governed by our genes, and how it is disrupted as we age. We will look at the consequences of that disruption for our cells and tissues and ultimately ourselves as individual beings. We will examine the fascinating question of why even though all living creatures are subject to the same laws of biology, some species live so much longer than even closely related ones, and what this might mean for us humans. We will take a dispassionate look at the most recent efforts being made to extend life span and whether they live up to their hype. I will also challenge some fashionable ideas, such as whether we do our best work in old age. I hope to probe, as well, the crucial ethical question that runs beneath anti-aging research: Even if we can, *should* we?

The first step in our journey is to think about what exactly death is, the many ways it can manifest itself, and explore the fundamental question of why we die.

1

THE IMMORTAL GENE

AND THE

DISPOSABLE BODY

Whenever I walk along the streets of London, I never cease to be amazed by a city where millions of people can work, travel, and socialize so seamlessly. A complex infrastructure, and hundreds of thousands of people, all work in concert to make it possible: the London Underground and buses to move us around the city; the post office and courier services to deliver the mail and goods; the supermarkets that supply us with food; the power companies that generate and distribute electricity; and the sanitation services that keep the city clean and remove the enormous quantities of waste we produce. As we go about our business, it is easy to take for granted this incredible feat of coordination that we call a civilized society.

The cell, our most basic form of life, has a similarly complex choreography. As the cell forms, it builds elaborate structures like the parts of a city. Thousands of synchronized processes are required to keep it functioning. It brings in nutrients and exports waste. Transporter molecules carry cargo from where they are made to distant parts of the cell where they are needed. Just as

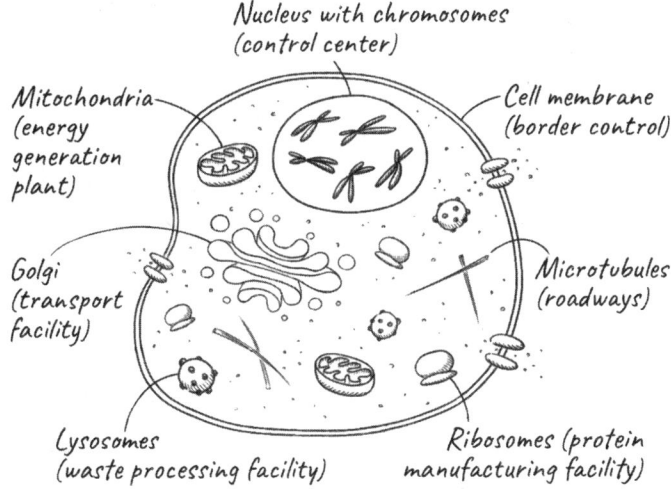

Nucleus with chromosomes
(control center)

Mitochondria
(energy
generation
plant)

Cell membrane
(border control)

Golgi
(transport
facility)

Microtubules
(roadways)

Lysosomes
(waste processing facility)

Ribosomes (protein
manufacturing facility)

The complex organization of a cell has similarities to a city. Only some of the major components are shown, and for clarity, they are not drawn to scale.

cities cannot exist in isolation but must exchange goods, services, and people with surrounding areas, the cells of a tissue need to communicate and cooperate with neighboring cells. Unlike cities, whose growth is not always constrained, the cell needs to know when to grow and divide but also when to stop doing so.

Throughout history, cities were imagined by their inhabitants to be permanent. We don't go about our lives thinking that the city we live in will one day cease to exist. Yet cities and entire societies, empires, and civilizations grow and die just as cells do. When we talk about death, we aren't usually thinking about these other kinds of death; we mean as it occurs to each one of us as individuals. But it turns out to be tricky even to define an individual, let alone what we mean by its birth or death.

At the moment of our death, what exactly is it that dies? At this point, most of the cells in our body are still alive. We can donate entire organs, and they work just fine in someone else if transplanted quickly enough. The trillions of bacteria, which outnumber the human cells in our body, continue to thrive. Sometimes the reverse is also true: suppose we were to lose a limb in an accident. The limb would certainly die, but we don't think of *ourselves* as dying as a result.

What we really mean when we say we die is that we stop functioning as a coherent whole. The collection of cells that forms our tissues and organs all communicate with one another to make us the sentient individuals we are. When they no longer work together as a unit, we die.

Death, in the inevitable sense we are considering in this book, is the result of aging. The simplest way to think of aging is that it is the accumulation of chemical damage to our molecules and cells over time. This damage diminishes our physical and mental capacity until we are unable to function coherently as an individual being—and then we die. I am reminded of the quote from Hemingway's *The Sun Also Rises*, in which a character is asked how he went bankrupt, and he replies, "Two ways. Gradually, then suddenly." Gradually, the slow decline of aging; suddenly, death. The process of aging can be thought of as starting gradually with small defects in the complex system that is our body; these lead to medium-sized ones that manifest as the morbidities of old age, leading eventually to the system-wide failure that is death.

Even then, it is hard to define exactly *when* this happens. Death used to mean when someone's heart stopped beating, but today cardiac arrest can often be reversed by CPR. The loss of brain function is now taken as a more direct sign of death, but there are hints that even that can sometimes be reversed. Differences in the precise legal definition of death can have very real consequences. Harvesting organs for donation from two persons in two different US states could be perfectly legal in one and murder in the other, even if they were both considered dead using identical criteria. A girl who was declared brain dead in Oakland, California, was considered alive by the standards of New Jersey, where her family lived. Her family petitioned and eventually had her body transported with its life support equipment to New Jersey, where she died a few years later.

If the precise moment of our death is ill-defined, so too is the moment of our birth. We exist before we emerge from the womb and take our first breath. Many religions consider conception to be the beginning of life, but conception too is a fuzzy term. Rather, there is a window of time after a sperm has made contact with the surface of an egg during which a series of events has to take place before the genetic program of the fertilized egg is set into motion. After that, there is a multiday window during which the fertilized egg undergoes a few divisions, and the embryo—now called a blastocyst—has to implant itself in the lining of the womb. Still later, the beginning of a heart develops, and only long after that, with the development of a nervous system and its brain, can the growing fetus sense pain.

The question of when life begins is as much a social and cultural question as it is a scientific one, as can be seen by the continuing debate over abortion. Even in many countries where abortion is legal, including the United States and the United Kingdom, it is a crime to grow embryos for research beyond fourteen days, which corresponds roughly to the time when a groove called the primitive streak appears in the embryo and defines the left and right halves. After this stage, the embryo can no longer split and develop into identical twins. Although we think of birth and death as instantaneous events—in one instant we come into existence and in another we cease to exist—the boundaries of life are blurry. The same is true of larger organizational units. It is hard to pinpoint the exact time when a city came into existence or when it crumbled.

Death can occur at every scale, from molecules to nations, but there are common features of the growth, aging, and demise of these very different entities. In every case, there is a critical moment when the component parts no longer allow the organic whole to function. Molecules in our cells work in a coordinated way to allow the cell to function, but they themselves can suffer chemical damage and eventually break down. If the molecules are involved in vital processes, their cells will themselves begin to age and die. Moving up the scale hierarchy, the trillions of cells in a human being carry out their specialized duties and communicate with one another to allow an individual to function. Cells in our body die all the time, with no adverse effects. In fact, during the growth of an embryo, many cells are programmed to die at precise points of development—a phenomenon called apoptosis. But when enough *essential* cells die, whether in the heart or the brain or some equally critical organ, then the individual can no longer function and dies.

We human beings are not so different from our cells. We carry out roles in groups: companies, cities, societies. The departure of one employee will not normally affect the functioning of a large company, and even less that of a city or a country, just as the death of a single tree says nothing at all about the viability of a forest. But if key employees, such as the entire senior management, were to leave suddenly, the health and future of the company would be in doubt.

It is also interesting to see that longevity increases with the size of the entity. Most of the cells in our body have died and been replaced many times before we ourselves die, while companies tend to have much shorter life spans than the cities in which they operate. The principle of safety in numbers has driven the evolution of both life and societies. Life probably began with self-replicating molecules, which then organized in closed compartments that we know as cells. Some of those cells then banded together to form individual animals. Then animals themselves organized into herds—or, in our case, communities, cities, and nations. Each level of organization brought greater safety and a more interdependent world. Today hardly any of us could survive on our own.

STILL, WHEN WE THINK OF DEATH, we are generally thinking about our own: the end of our conscious existence as an individual. There is a stark paradox about that kind of death: although individuals die, life itself continues. I don't mean just in the sense that our family, community, and society will all go on without us. Rather, it is remarkable that every creature alive today is a direct descendant of an ancestral cell that existed billions of years ago. So, although changing and evolving with time, some essence in all of us has lived continuously for a few billion years. That will continue to be true for every living thing for as long as life survives on Earth, unless we one day create an entirely artificial form of life.

If there is a direct line of succession from us to our ancient ancestors, then there must be something about each of us that *doesn't* die. That something is information on how to create another cell or an entirely new organism, even after the original carrier of that information has died—just as the ideas and information here can persist in some form long after the physical copy of this book has deteriorated.

The information to continue life resides, of course, in our genes. Each gene is a section of our DNA, and is stored in the form of chromosomes in the nucleus, the specialized compartment that encapsulates genetic material in our cells. Most of our cells contain the same entire set of genes, known collectively as our genome. Every time our cells divide, they pass on the entire genome to each of the daughter cells. The vast majority of these cells are simply part of our body and will die with it. But some of our cells will outlive our body by developing into our children—the new individuals that make up the next generation. So what is special about these cells that allows them to live on?

The answer to this settled a raging controversy, one that came long before our knowledge of genes, let alone DNA. When people first began to accept that species could evolve, two opposing views emerged. The first, advanced by the Frenchman Jean-Baptiste Lamarck in the early nineteenth century, held that acquired characteristics could be inherited. For example, if a giraffe were to keep stretching its neck to reach higher branches for leaves to eat, its offspring would inherit the resulting longer neck. The second theory was natural selection, proposed by a pair of British biologists, Charles Darwin and Alfred Wallace. In this view, giraffes were variable, some with longer necks and others with shorter. Those with longer necks were more likely to find nourishment and thus be able to survive and have offspring. Progressively, with each generation, variants with longer and longer necks would be selected.

A relative outsider working in what was then the Malay Archipelago, thirty-five-year-old Alfred Wallace wrote to Darwin in 1858 expressing his ideas, not realizing that the older man had himself come to the same conclusion many years earlier. Because these ideas were so revolutionary, and had social and religious implications, Darwin had not yet summoned the courage to publish them, but the communication from Wallace spurred him into action. Darwin was at the heart of the British scientific establishment, and had he been less scrupulous, he could have simply ignored Wallace's letter and hurriedly published his book. Nobody would have ever known Wallace's name. Instead, Darwin arranged for himself and Wallace to make a joint presentation at the Linnean Society of London on July 1, 1858. The response to the lecture itself was relatively muted and had little immediate impact. In what was one of the worst pronouncements in the history of science, the society's president said in his annual address, "The year has not, indeed, been marked by any of those striking discoveries which at once revolutionize, so to speak, the department of science on which they bear." However, the lecture paved the way for the publication of Darwin's book *On the Origin of Species* the following year, which changed our understanding of biology forever.

In 1892, thirty-three years after Darwin's monumental tract was published, the German biologist August Weismann posited a neat rebuttal of Lamarck's ideas. Although humans have known for a very long time that sex and procreation were connected, it is only in the last 300 years that we discovered that the key event is the fusion of a sperm with an egg to start the process. The fertilization of an egg by a sperm results in the seemingly miraculous creation of an entirely new individual. The individual consists of trillions of cells that carry out nearly all of the functions of the body and die with it. They are known collectively as somatic cells, from *soma*, the Latin and Greek word for "body." The sperm and the egg, on the other hand, are germ-line cells. They reside in our gonads, which are testes in males and ovaries in females. And they are the sole transmitters of heritable information: our genes. Weismann proposed that germ-line cells can create the somatic cells of the next generation, but the reverse can never happen. This separation between the two kinds of cells is called the Weismann barrier. So if a giraffe stretches its neck, it might affect various somatic cells that make up its neck muscles and skin, but these cells would be incapable of passing on any changes to its offspring. The germ-line cells, protected in the gonads, would be impervious to the activities of the giraffe and any characteristics its neck acquired.

The germ-line cells that propagate our genes are immortal in the sense that a tiny fraction of them are used to create the next generation of both somatic and germ-line cells by sexual reproduction, which effectively resets the aging clock. In each generation, our bodies, or our soma, are simply vessels to facilitate the propagation of our genes, and they become dispensable once they have fulfilled their purpose. The death of an animal or a human is really the death of the vessel.

WHY DOES DEATH EVEN EXIST? Why don't we simply live forever?

The twentieth-century Russian geneticist Theodosius Dobzhansky once wrote, "Nothing in biology makes sense except in the light of evolution." In biology, the ultimate answer to a question about *why* something occurs is because it evolved that way. When I first began to consider the question of why we die, I thought naively that perhaps death was nature's way of allowing a new generation to flourish and reproduce without having older ones hanging around to compete with it for resources, thus better ensuring the survival of the genes. Moreover, each member of a new generation would have a different combination of genes than its parents, and the constant reshuffling of life's deck of cards would help facilitate survival of the species as a whole.

This idea has existed at least since the Roman poet Lucretius, who lived in the first century BCE. It is appealing—but it's also wrong. The problem is that any genes that benefit the group at the expense of the individual cannot be stably maintained in the population because of the problem of cheaters. In evolution, a "cheater" is any mutation that benefits the individual at the expense of the group. For example, let us suppose there are genes that promote aging to ensure that people die off in a timely way to benefit the group. If an individual had a mutation that inactivated those genes and lived longer, that person would have more opportunity to have offspring, even though it did not benefit the group. In the end, the mutation would win out.

Unlike humans, many insects and most grain crops reproduce only once. Species such as the soil worm *Caenorhabditis elegans*, as well as salmon, produce lots of offspring in one big bang and die in the process, often recycling their own bodies as a form of suicide. This kind of reproductive behavior makes sense for worms, which usually live as inbred clones and are therefore genetically identical to their offspring. On the other hand, the reproductive behavior of salmon is a result of their life cycle: they have to swim thousands of miles in the ocean before returning to spawn. With little chance of surviving such a journey twice, they are better served by putting everything they can into breeding just once, using up their entire energy and even dying in the process, to produce enough offspring and maximize the chance that those offspring survive. For species that can reproduce multiple times, like humans, flies, or mice, it would not make genetic sense to die in the act of producing off-spring to which they are only 50 percent related. In general, natural selection rarely acts for the good of species or even groups. Rather, nature selects for what evolutionary biologists call fitness, or the ability of individuals to propagate their genes.

If the goal is to ensure that our genes are passed on, why has evolution not prevented aging in the first place? Surely the longer humans survive, the more chance we have of producing offspring. The short answer is that through most of our history as a species, our lives were short. We were generally killed by an accident, disease, predator, or a fellow human before our thirtieth birthday. So there was no reason for evolution to have selected us for longevity. But now that we have made the world safer and healthier for us, why don't we just keep living on?

The solution to this puzzle began in the 1930s with two members of the British scientific elite, J. B. S. Haldane and Ronald Fisher. Haldane was a polymath who worked on everything from the mechanisms of enzymes to the origin of life. He was a socialist who late in life became disillusioned with Britain and emigrated to India, where he died. Fisher's fundamental contributions to statistics have propelled our understanding of evolution and also form the basis of randomized clinical trials that are used to test the efficacy of new drugs or medical procedures and have saved millions of lives. More than fifty years after his death in 1962, he became controversial for his views on eugenics and race. A stained glass window that portrayed one of Fisher's key ideas for the design of experiments was recently removed by Gonville and Caius College in Cambridge, where he was once a fellow, and its final disposition is still uncertain.

Around the same time, Fisher and Haldane independently came up with a revolutionary idea. A mutation that is harmful early in life, each realized, would be strongly selected against because those who carry it would not reproduce. However, the same could not be said for a gene that is deleterious to us only *later* in life, because by the time it causes harm, we will already have passed it on. For most of our history as a species, we would not have even noticed its harmful consequences, because long before these effects would be felt, we would have died. It is only relatively recently that we have become aware of the consequences of any mutations that are detrimental late in life. Huntington's disease, for example, primarily affects people over thirty, by which time, historically, most of them would have already reproduced and died.

Fisher's and Haldane's ideas explain why certain deleterious genes persist in the human population, but their relevance to aging was not immediately obvious. That understanding came when British biologist Peter Medawar, another brilliant and colorful figure, turned his attention to the problem. Medawar, born in Brazil, was most famous for his ideas of how the immune system rejects organ transplants and acquires tolerance. Unlike many scientists who focus narrowly on one area, Medawar, like Haldane, had widespread interests, and wrote books that were famous for their erudition and elegant writing. Many scientists of my generation grew up reading his *Advice to a Young Scientist* (1981), which I found pompous, arrogant, thoughtful, engaging, and witty all at once.

Medawar proposed what has become known as the mutation accumulation theory of aging. Even if a person harbored multiple genetic mutations that didn't noticeably impair health early on, in combination they brought about chronic problems later in life, resulting in aging.

Going one step further, the biologist George Williams suggested that aging occurs because nature selects for genetic variants, even if they are deleterious later in life, because they are beneficial at an earlier stage. This theory is called antagonistic pleiotropy. *Pleiotropy* is simply a fancy term for a situation in which a gene can exert multiple effects. So antagonistic pleiotropy means that the same gene could have opposite effects; with genes involved in aging, the effects could occur at different times, such as being helpful early in life and problematic later. For example, genes that help us grow early in life increase the risk of age-related diseases such as cancer and dementia when we are old.

Similarly, the disposable soma hypothesis posits that an organism with limited resources must apportion them between investing in early growth and reproduction and prolonging life by continuously repairing wear and tear in the cell. According to biologist Thomas Kirkwood, who first proposed this theory in the 1970s, the aging of an organism is an evolutionary trade-off between longevity and increased chances of passing on its genes through reproductive success.

Is there any evidence for these various ideas about aging? Scientists have experimented on fruit flies and worms, two favorite organisms because they are easy to grow in the laboratory and have short generation times. Exactly as these theories would predict, mutations that increase life span reduce fecundity (the rate at which an organism produces offspring). Similarly, reducing the caloric intake of the daily food given to these organisms also increases life span and reduces fecundity.

Apart from the ethics of experimenting on humans, the two to three decades between generations is too long for a typical academic career, let alone the handful of years a graduate student or research fellow might stick around. But an unusual analysis of British aristocrats over the past 1,200 years shows that among women who survived beyond sixty (to weed out factors such as disease, accidents, and dying in childbirth), those with fewer children lived the longest. The authors argue that in humans too, there is an inverse relationship between fecundity and longevity, although, of course, as any harried parent knows, there could have been many other reasons why having fewer children extends life expectancy.

THE INCREASE IN OUR LIFE span over the last century brings us to another curious feature of aging that is almost unique to humans: menopause. With the exception of a few other species, including killer whales, most female animals can reproduce almost to the end of their lives, whereas women suddenly lose the ability in midlife. The abruptness of this change in women, as opposed to the more gradual decline in male fertility, is also strange.

You might think that if evolution selects for our ability to pass on our genes, it should want us to reproduce for as much of our lives as possible. So why do women stop reproducing relatively early in life?

This may be asking the wrong question. Our closest relatives, such as the great apes, all stop having babies about the same age that we do: the late thirties. The difference is that they generally die soon afterward. And for most of human history, most women too died soon after menopause, if not earlier. Perhaps the real question is not why menopause occurs so early in life but why women live so long afterward.

People cannot be sure they have reproduced in the sense of passing on their genes until their youngest child has become self-sufficient, and humans have a particularly long childhood during which they are dependent on their parents. Menopause may have arisen to protect women from the increased risk of childbirth in later age, keeping them alive longer to take care of the children they had already. This might also explain why men—who don't suffer such an increased risk—can be reproductive until much later in life. So perhaps menopause developed as an adaptation to maximize the chances of a woman's children growing up—and thus propagating her genes. This is the so-called good mother hypothesis. Indeed, the few species where females live well beyond their reproductive years are ones whose offspring require extended maternal care. However, even in these species, there is a gradual loss of fertility rather than the abrupt change brought on by menopause. For example, although the fertility of elephants declines with age, they, unlike humans, can continue to have offspring until very late in life. Similarly, while living beyond childbearing age has also been observed in chimpanzees, menopause actually occurs near the end of their life span.

The grandmother hypothesis for the origin of menopause takes the idea one generation further. Proposed by the anthropologist Kristen Hawkes, it argues that living longer makes sense if a woman helps in the care of her grandchildren, thus improving their survival and ability to reproduce. But others contend that it is rarely better for a woman to give up the chance to pass on half her genes through continuing to have her own children for the sake of improving the survival of grandchildren, who only carry a quarter of her genes.

Another idea, based on studying killer whales, one of the few species that, like humans, has true menopause and lives in groups, is that menopause is a way to avoid intergenerational conflict. In some species that breed in groups, reproduction is suppressed in younger females, who act as helpers to older, reproducing females. But in humans, there is little overlap: women stop breeding when the next generation starts to breed. Women would have no interest in helping their mother-in-law have more children, since they would not have any genes in common. But a woman who helps her daughter-in-law reproduce will help to bequeath a quarter of her genes to her grand-children. So her best strategy may be to stop breeding and help her daughter-in-law breed instead.

It could also simply be that the number of eggs in a female evolved to match its average life span in the wild. Steven Austad, now at the University of Alabama in Birmingham, points out that menopause may not be adaptive at all in the sense of favoring mothering or grandmothering. It was only about forty thousand years ago that we became much longer lived than Neanderthals and chimpanzees. So perhaps there has just not been enough time for the aging of human ovaries to adapt to that increased life span. In the absence of hard experiments, scientists, especially evolutionary biologists, love to argue.

THESE THEORIES OF WHY WE age depend on the idea of a dispos-able body being able to pass on its genes before it ages and dies. In doing so, the aging clock is somehow reset with each generation. Such theories should apply only to organisms where there is a clear distinction between parents and offspring. Certainly that distinc-tion is true for all sexual reproduction. Sex evolved because it is an efficient mechanism to produce genetic variation in the offspring by generating different combinations of genes from each parent, allowing organisms to adapt to changing environments. In some sense, you could say that death is the price we pay for sex! While this may be a catchy statement, not all animals with a distinction between germ line and soma reproduce sexually. Moreover, scientists have found that even single-celled organisms such as yeast and bacteria age and die, as long as there is a clear distinction between mother and daughter cells.

The laws of evolution apply to all species, and all life forms are made up of the same substances. Biologists from Darwin onward have never ceased to be amazed that evolution, which is simply selecting for fitness—or the efficiency with which each species can pass on its genes—has given rise to the amazing variety of life forms on Earth. That variety includes a huge range of life spans, from those best measured in hours to those that may stretch more than a century. For human beings seeking to understand the potential limits of our own longevity, some surprising lessons can be learned from species across the animal kingdom.

2.

LIVE FAST
AND DIE YOUNG

In springtime, my wife and I will often take a walk in Hardwick Wood near Cambridge to see the riot of bluebells that cover the forest ground. Once, we were walking along a path when we came upon a stone monument commemorating Oliver John Hardiment, a young man who died in 2006 at the age of twenty-five. Below his name was a quotation from the Indian writer Rabindranath Tagore: "The butterfly counts not months but moments and has time enough."

The life of a butterfly can be as short as a week, and most live less than a month. As I considered the fleetingly short life of a typical butterfly, I was reminded of the contrast with something else that had fascinated me. I have often visited the American Museum of Natural History in New York, where there is an enormous section of the trunk of a giant sequoia tree. The tree was more than 1,300 years old when it was cut down in 1891. Some yew trees in Britain are estimated to be over 3,000 years old.

Of course, trees are fundamentally different from us because of their ability to regenerate. In the Cambridge University Botanic Garden there is an apple tree that was grown from a cutting from the tree under which a young Isaac Newton sat a few hundred years ago about a hundred miles north at Woolsthorpe Manor, the Newton family home. In fact, there are several "Newton" trees, all started as cuttings from the one with the famous apple that fell to the ground, allegedly inspiring Newton to formulate the theory of gravity. The question of whether these trees should be dated back to the root system of the original is interesting, but it is different from looking at the life span of animals.

Even in the animal kingdom, there are some species that possess tree-like properties. If you cut off one of a starfish's arms, it can grow right back. A small aquatic animal called a hydra is even more impressive: it doesn't seem to age at all and is able to regenerate tissue continuously. Still, it is a complex procedure. One study showed that a large number of genes are involved just for regenerating its head. All this for an organism that is barely half an inch in length.

If the hydra is remarkable, it is related to another sea dweller that can age backward—at least metaphorically. That species is *Turritopsis dohrnii*, also known as the immortal jellyfish. This jellyfish, when faced with injury or stress, will metamorphose into an earlier stage of development and live its life all over again. It is almost as if an injured butterfly could transform itself back into a caterpillar and start over.

Since hydra and the immortal jellyfish don't exhibit obvious signs of aging, they are often called biologically immortal. This doesn't mean they don't die—they can and do die for all sorts of reasons. They still fear predators and must themselves obtain enough food to survive. Nor does it even mean that they *cannot* die of biological causes. But, unlike most every animal, their likelihood of dying does not increase with age.

Species such as hydra and the immortal jellyfish excite gerontologists because they may provide clues about how to defeat the aging process. But to me, their property of being able to regenerate entire body parts, or even a whole organism, makes them more similar to trees than to us. Although we may learn some fascinating things about their lack of apparent aging, it is not at all clear how relevant those findings will be to human aging. Sometimes biology is universal, especially if it relates to fundamental mechanisms. But in other cases, even discoveries in rats or mice, which are mammals and biologically much closer to us, are difficult to translate into humans. It may be a very long time before any findings gleaned from hydra or jellyfish are useful to us.

PERHAPS WE NEED TO LOOK at species that are more closely related to us—say, mammals, or at least vertebrates. Although this class of animals doesn't span the enormous range of longevity from insects to trees, they still vary considerably. Some small fish live for just a few months, while a bowhead whale is known to have lived for more than 200 years, and a Greenland shark is thought to have lived almost *400* years.

What causes this large variation even among a particular group of animals such as mammals? Can we detect a pattern among these species just from some overall characteristics? Scientists have long looked for such relationships. Physicists, especially, love to look for general rules to make sense of disparate observations. Geoffrey West at the Santa Fe Institute is one such physicist who now works on complex systems, including aging. West takes a broad view, analyzing how cities and companies, as well as organisms, grow, age, and die. Along the way, he explores how some properties of animals scale across a wide range of sizes and longevities.

If you look at mammals, the larger the animal, generally speaking, the longer its life span. This makes evolutionary sense. A small animal is more vulnerable to predators, and there would be no point in having a long life span if it is going to be eaten long before it dies of old age. But the more fundamental reason for the relationship between size and life span is that size is related to metabolic rate, which is roughly the rate at which an animal burns fuel in the form of food to provide the energy it needs to function. Small mammals have more surface area for their size and so lose heat more easily. To compensate, they need to generate more heat, which means maintaining a higher metabolic rate and eating more for their weight. This means that the total number of calories burned per hour by an animal increases less slowly than the mass of the animal. An animal that is ten times as large burns only four to five times as many calories per hour. So for their weight, smaller animals burn more calories than larger animals. The relationship between how fast an animal burns calories and its mass is named Kleiber's law after Max Kleiber, who showed in the 1930s that an animal's metabolic rate scales to the $3/4^{th}$ power of its mass. The exact power is a matter of dispute and some show that for mammals, a $2/3^{rd}$ power fits the data better.

Since heart rate also scales with metabolic rate, over a very wide range of sizes—from hamsters to whales—mammals typically have roughly the same number of heartbeats over their lifetime: about 1.5 billion. Humans currently have almost twice that, but, then, our life expectancy has doubled over the last hundred years. It is almost as if mammals were designed to last a certain number of heartbeats, much like a typical car can be driven about 150,000 miles. West points out that 1.5 billion is also roughly the number of total revolutions a car engine makes over its expected lifetime and asks, perhaps tongue in cheek, whether this is just a coincidence or whether it tells us something about the common mechanisms of aging!

These relationships suggest that there will be natural limits on life span because size and metabolic rate can vary only so much. For example, an animal cannot evolve to become arbitrarily large without collapsing under its own weight. Such an animal would also have great difficulty supplying its cells with the necessary oxygen. A metabolism must be fast enough for an animal to move and find food—and there are biological limits on how fast a metabolism is actually achievable if you are small. But within the allowable range, these rules hold remarkably well. Geoffrey West declares that just knowing the size of a mammal, he could use scaling laws to estimate almost everything about it: from its food consumption, to its heart rate, to its life span.

This is quite remarkable, and although it deals with averages, it sounds almost like a hard-and-fast rule that limits life span. But what of human beings' marked increase in longevity over the past century? As West observes, this is a question of what one means by life span: we have almost doubled life expectancy in the last hundred years, but we have done nothing at all to increase the *maximum* human life span, which remains about 120 years. He argues that, according to the evidence, aging and mortality result from the wear and tear of being alive. Inexorable forces of entropy—a measure of disorder—that push in the direction of disorder and disintegration press against that dream of immortality. Unlike cars, which consist of mechanical components that we can swap out for new ones as they wear out, we cannot simply replace ourselves with new parts and keep going indefinitely.

WHILE THIS RULE-OF-THUMB CONNECTION AMONG size, metabolism, and life span is fascinating, biologists tend to be more interested in the exceptions. They love to study species that beat the system, in the hopes that they can tell us something about the underlying mechanisms of aging. One big question is whether there is a theoretical maximum life span or not. We have seen species such as hydra and jellyfish that seem not to age and can, in fact, continuously replace their worn-out parts. While biologists are well aware of the second law of thermodynamics—which states that in any natural process the amount of disorder or entropy increases with time—most would disagree that the law applies in some blanket form to aging and death, because living systems are not closed as the law requires but need a constant input of energy to exist. In fact, with a sufficient expenditure of energy, you can indeed reverse entropy when it comes to regularly cleaning your attic or hard drive; it is just that most of us don't feel it is worth it.

As a result, biologists do not think that aging is inevitable. Rather, all evolution cares about is fitness: the ability to pass on our genes most efficiently. But living a long life is worth it only if you are not going to be eaten or die of disease or an accident long before you die of old age. Hence birds, which can escape predators by flying away, generally live longer than earthbound animals of about the same size. For those lucky animals that don't have as much to fear from predators, living a longer life gives them more time to find a mate and reproduce. Slowing down their metabolism, so that they need not procure large amounts of food every day, may then simply be a way of surviving better into old age. In each case, the life span simply reflects how evolution has optimized the fitness of each species.

Steven Austad is a leader in aging research who studies exotic species with widely varying life spans. For a scientist, he has a highly unusual background: he majored in English literature at the University of California, Los Angeles, hoping to write the Great American Novel. Given that we've never heard of it, Austad jokes, one can see how that worked out. After graduation, while not writing his novel, he drove a taxi and worked as a newspaper reporter before spending several years taming lions, tigers, and other wild animals for the movie industry. This sparked an interest in science, and Austad went back to school to study animal behavior. From there, he became interested in the question of why animals age at different rates.

In 1991 Austad and his graduate student Kathleen Fischer examined the longevity of several hundred species. They discovered that, even among mammals, the relationship between body size and longevity disappears below a threshold of about one kilogram of body mass. Possessing a biologist's instinct for the particular, the two of them then asked which species deviated most from this scaling law, coining what they called the longevity quotient. The LQ is the ratio of the average life span of the species to what it would be if it followed the scaling laws. This allowed them to focus on those species that deviate by either living much longer or much less than would be expected for their size.

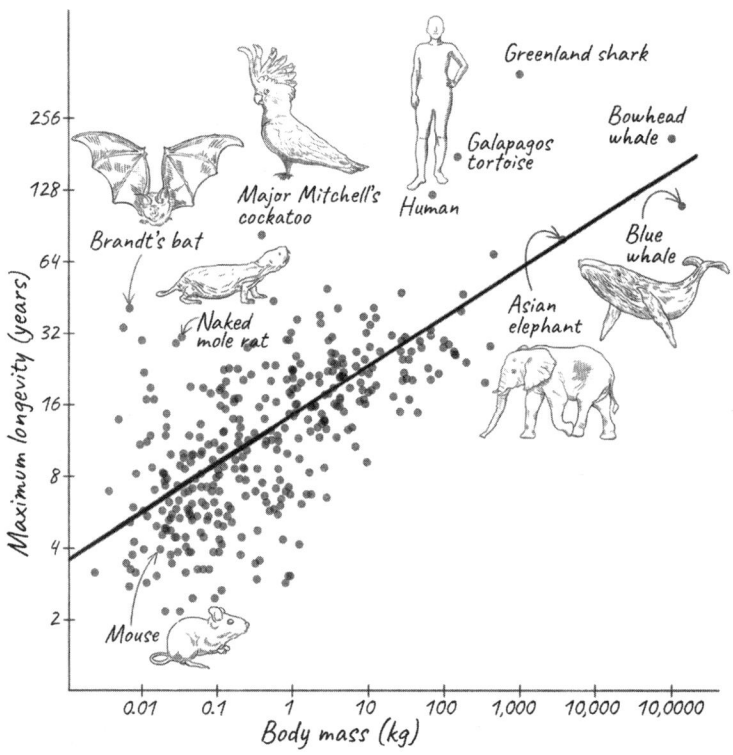

The life span of animals generally increases with size. Estimates for the maximum life span of mammals are shown along with a line showing the general trend. In addition, points for the Major Mitchell's cockatoo, Galapagos tortoise, and Greenland shark are shown. Data are taken from the AnAge database (https://genomics.senescence.info/species/index.html).

It turns out that humans already do rather well: we have an LQ of about 5, meaning that we live 5 times as long as would be expected. Nineteen mammalian species outperform us: eighteen species of bat and the naked mole rat. Over the years, Austad has studied these outlier species, and he describes them in colorful prose as befits his background in English literature. He poses this provocative question: Why do aging researchers study mice and rats, both of which have LQs of just 0.7, when they could be looking at these more exceptional species instead? There are many reasons why animals are chosen as model organisms, including ease of breeding and maintenance, and the ability to study their genetics. We have acquired tremendous knowledge of their biology over decades. Since the underlying mechanisms of aging are likely to be universal even if their rates are not, and studying short-lived animals could actually be an advantage by speeding up experiments, I am not sure that many in the gerontology community will rush to follow Austad's advice. But I hope enough of them do, so that we learn how these unusually long-lived outliers have evolved such different rates of aging.

Among the species Austad describes are giant tortoises, such as the Galápagos tortoise, which holds the record for life span of a terrestrial vertebrate animal and can amble along for two centuries. There might well be a Galápagos tortoise still alive that was spotted by Darwin during his five-year voyage aboard the Royal Navy ship HMS *Beagle* from 1831 to 1836. Also, for much of their long life, they are remarkably free of diseases such as cancer. Determining the LQ of these tortoises is tricky, though. For one, their exact age is hard to determine, since their history is usually poorly documented and the subject of much exaggeration. Even thornier is the question of what a tortoise truly weighs. Much of their body mass consists of their protective shell, which is more like our hair and nails than highly active tissue, so drawing comparisons with other animals can be misleading.

These giant tortoises may not be alone in their longevity. Two studies that evaluated survival data from various turtles and other reptiles and amphibians found negligible senescence in a number of turtles and other species. The biologist's term *negligible senescence*, which means little or no increase in mortality, has been interpreted popularly to mean "eternal life," but this is a bit of a misnomer. Actually, it means that mortality, or the likelihood of dying, does not increase with age.

The relationship between mortality and age was worked out in 1825 by Benjamin Gompertz, a self-educated British mathematician. Gompertz worked for an insurance company, and so was naturally interested in the question of when a person seeking to purchase coverage might die. By digging through death records, he discovered that starting in our late twenties, the risk of dying increases at an exponential rate year after year. It doubles roughly every seven years. At age 25, our probability of dying in the next year is only about 0.1 percent. This rises to 1 percent at age 60, 6 percent at age eighty, and 16 percent at age 100. By the time a person reaches 108 years old, there is only about a 50 percent chance of making it another year.

Negligible senescence, when the probability of dying is constant rather than exponentially increasing with age, violates Gompertz's law. But even if there is negligible or even negative senescence, you still face a probability of dying every year from age-related diseases, quite apart from dying of infections or accidents. Aging involves more than increasing mortality with age. It also depends on maintaining the physiology of the animal. The long-lived tortoises show unmistakable signs of aging. Like elderly humans, their eyesight and heart gradually fail. Some of them develop cataracts. Some become feeble to the point where they need to be fed by hand. So these animals *do* age, just slowly.

Moreover, biological time for tortoises is very different: they live life in the slow lane. They are not warm-blooded creatures like us mammals. They move slowly and reproduce slowly, often taking several decades to reach puberty in the wild. Their hearts beat only once every ten seconds, and they breathe slowly. Despite their long chronological lives, they fit the metabolic rate theory of longevity.

Other long-lived species are aquatic, such as the Beluga sturgeon and the aforementioned Greenland shark. Like the tortoise, they too aren't in any hurry. Greenland sharks swim more slowly than a normal eighty-year-old human walks, and they seem to be scavengers, rather than catching prey. Perhaps more extraordinary than the Greenland shark is the bowhead whale. This baleen whale lives in freezing Arctic waters, but because it is a warm-blooded mammal, its internal body temperature is only a few degrees lower than that of most other mammals. Moreover, it eats about three times more than was previously suspected, implying a metabolic rate three times higher than was thought. How such an animal can survive for about 250 years is still a mystery.

The Greenland shark and the bowhead whale are large aquatic vertebrates, but there are much smaller terrestrial outliers too. One particularly interesting example is Major Mitchell's cockatoo, a striking white bird with a pink face and a vibrant bright red and yellow crest that resembles a radiating sun. This cockatoo has been known to live to eighty-three years in a zoo. This would not be exceptional for a human, but the bird is far smaller. So this is definitely not a species that fits the general relationship among size, metabolic rate, and life span.

Remember how the relationship between mass and longevity for mammals disappeared below one kilogram? That's largely due to bats. Bats do not live as long as Major Mitchell's cockatoo, but they generally outlive nonflying mammals of the same size, which is exactly what evolutionary theories would predict, since their ability to fly allows them to evade predators. In keeping with this, bats that roost in caves, and are thus further protected from predators, live almost five years longer than those that don't. The champion is Brandt's bat, a small, brown animal that fits comfortably in the palm of your hand. A male of the species was recaptured in the wild forty-one years after it was originally banded. Austad estimates that its LQ of about 10 is the highest known for any mammal and about twice that of humans.

Another reason bats are thought to live longer is that they slow down their metabolism during their long periods of hibernation. On average, bats that hibernate live six years longer than those that don't. But even bats that don't hibernate live exceptionally long for their size, so clearly metabolic rate is not the only reason for their longevity. Rather, they may have special mechanisms that protect them from aging.

One curious feature is that the longest-lived Brandt's bats on record are males. This is certainly different from humans. Austad speculates that this could be because female bats are less agile in flight and more susceptible to predators when they are pregnant, because they carry more than a quarter of their own body weight. They also face much greater energy demands in feeding their young.

Finally, no discussion of long-lived animals would be complete without mentioning the remarkably ugly, nearly hairless rodent that has become something of a darling of the aging research community: the naked mole rat. Despite the name, it is neither a mole nor a rat but a species of rodent that is indigenous to equatorial East Africa. It is about the same size as a mouse, but whereas a mouse lives roughly two years, a naked mole rat can live for more than thirty. This gives it an LQ of 6.7—not as high as Brandt's bat, but a record for a terrestrial nonflying mammal. How do they do it?

Rochelle Buffenstein, currently at the University of Illinois in Chicago, has done more than perhaps anyone else to understand the biology of aging in the naked mole rat. As a result of work by her and many others, we know that naked mole rats are one of a small number of mammals that are referred to as eusocial: they live in underground colonies with a queen, and, in that sense, are reminiscent of ants. As one might expect, they have a very low metabolism and are tolerant of oxygen levels so low that they would kill mice— and us. In the wild, naked mole rat queens live much longer than workers: about seventeen years compared with two to three years. But in the lab, where worker naked mole rats live a comfortable, well-fed life with good health care and no predators, the difference is not so stark.

Not surprisingly, naked mole rats are extremely resistant to cancer, regardless of age—again, in marked contrast to mice. Even more strikingly, when Buffenstein and her colleagues tried to induce cancer in naked mole rat skin cells using techniques that worked reliably for other species, they could not do it. According to their 2010 study, instead of proliferating like cancerous cells, the naked mole rat cells entered a terminal state and were cleared away, suggesting that they respond to cancer-causing genes very differently.

One of the biggest headlines about naked mole rats was generated by the observation that they seem to violate Gompertz's law: their risk of dying seems not to increase with age. As a result of these findings, no animal has been hyped as much as the naked mole rat, with both the popular press and news articles in scientific journals touting each discovery as a major breakthrough in the quest to defeat aging. This was too much for some scientists, who pointed out that naked mole rats *do* age, just more slowly than might be expected for their size. As we saw with long-lived tortoises, they show many signs of aging, including lighter, thinner, and less elastic skin resembling parchment, as well as muscle loss and cataracts. They are not like hydra and the immortal jellyfish, which can regenerate themselves with ease. Still, as exceptionally long-lived mammals, they could provide important clues into our own aging processes.

IT IS TIME TO LEAVE these unusually long-lived species and focus on the one that interests us most: ourselves. Most crucially: How long can human beings live? And is this limit fixed, or can it be changed?

For most of human history, life expectancy was just over thirty. But today, in developed countries, we can look forward to living into our mid-eighties. Even in poorer countries, a person born today can expect to live longer than the grandparents of people in the richest countries. The science writer Steven Johnson makes the point that this is like each of us acquiring an entire additional life.

When we say life expectancy, we mean life expectancy at birth, or the average number of years a newborn would live if current mortality rates remained unchanged. This value, as you can imagine, is greatly affected by infant mortality rates. Even in the nineteenth century, when life expectancy was forty years, a person who reached adulthood had a good chance of living to be sixty or more. Most of the increase in life expectancy has come about because of improvements in public health rather than groundbreaking advances in medicine. Johnson observes that the three biggest contributors have been modern sanitation and vaccines, which both prevented the spread of infection, and artificial fertilizers. Other significant innovations were antibiotics, blood transfusions (crucial for accidents and surgery), and sterilization of water and food by chlorination and pasteurization.

The inclusion of fertilizers may surprise you, but prior to the ready availability of food—which has brought about its own problems of obesity, diabetes, and cardiovascular diseases—humans were constantly struggling to get enough to eat. Chemical fertilizers include nitrogen-containing compounds and have increased crop yields several-fold. The ability to chemically capture nitrogen from the air, a discovery for which Fritz Haber received the Nobel Prize in 1918, made it much easier to synthesize fertilizers and helped to double the world's population. Interestingly, almost half of the nitrogen atoms in our bodies went through a Haber-Bosch high-pressure steam chamber that converted atmospheric nitrogen to ammonia for use in fertilizers, which then ended up in the food we ate and became incorporated into ourselves.

Haber himself was a tragic figure. A German Jew, he was intensely loyal to Germany during World War I, and his method for fixing nitrogen into ammonia enabled the country to prolong the war by producing its own explosives. Prior to that, its military had been importing nitrates from Chile, which became impossible due to the Allied Powers' wartime blockade. He also initiated the use of chemical warfare against the Allies, who denounced him as a war criminal. At the same time, his Jewishness trumped his loyalty to Germany. Soon after the Nazis assumed power, he had to flee Germany in 1933 although he was a world-famous scientist and director of a prestigious institute in Berlin. After a brief sojourn in England, he set out for Rehovot in what is now Israel, but died mid-journey of heart failure in a hotel in Basel, Switzerland.

Back to life expectancy: preventing infectious disease dramatically reduced infant mortality, which is now as low as 1 percent in advanced countries and about 3–4 percent worldwide. But there has been progress across the rest of the aging curve as well. Public health measures for safety, regulations against smoking, and better treatments for life-threatening illnesses such as cardiovascular disease and cancer have all added up to a slow but steady increase in life expectancy beyond sixty years of age. Does this mean that our life expectancy might go on increasing indefinitely?

Ever since humans became aware of their mortality, we have wondered whether our life span has a fixed limit. Scientists aren't sure.

Jay Olshansky of the University of Illinois at Chicago says yes. He examined how much we would gain by eliminating various common causes of death such as cancer, heart disease, and other diseases. Based on statistical calculations, he argued that for life expectancy to increase dramatically, we would need to reduce mortality rates from all causes by 55 percent and even more at older ages. He and his colleagues contended that average life expectancy would likely not exceed eighty-five and that it would not exceed a hundred until everyone alive today had died. Even curing all forms of cancer would add only four to five years on average.

In the other corner was the late James Vaupel, who maintained that life span is elastic. If evolutionary theories were strictly correct, then our maximum life span should be adapted for life in the wild and thus not much more than about thirty to forty years. But, as you know, life expectancy has more than doubled. Moreover, in certain species, such as some tortoises, reptiles, and fish, mortality actually falls and then levels off, presumably because as these creatures grow larger, they can better resist starvation, predators, and disease; senescence is not inevitable.

The disagreements between the two boiled into a sort of scientific blood feud, with Vaupel refusing to attend any meetings where Olshansky was present, and attacking his findings as a "pernicious belief sustained by ex-cathedra pronouncements." Olshansky, for his part, feels that demographers relying purely on statistics fail to consider biology. In agreement with this, an analysis of the lives of primates implies that there are biological constraints on how much the rate of human aging can be slowed.

Of course, life expectancy at birth is not the same as the maximum possible life span, and it is that maximum that tends to interest us more than averages. We want to know how long it is theoretically possible for humans to live. Most cultures have writings about prophets and sages who allegedly lived for hundreds of years. In Western culture, the name Methuselah has become synonymous with longevity, after the biblical prophet who is said to have lived 800 years. In somewhat more recent times, the Englishman Tom Parr, who died in 1635, was said to have lived for 152 years, but this has been thoroughly debunked. Unlike most people, for whom childhood memories are the strongest, "Old Tom" could remember nothing of his youth.

The oldest person for whom we have reliable records is Jeanne Calment, who died at the age of 122 in 1997. She lived in Arles, the town in southern France where van Gogh resided near the end of his life. She actually met the troubled artist in her teens, describing him as "very ugly, ungracious, impolite, and sick." Apparently Calment had a sharp wit. As she grew older and older, journalists began to gather around her on each birthday. When one of them took leave by telling her, "Until next year, perhaps," she retorted, "I don't see why not! You don't look so bad to me."

Calment was in very good health for nearly her entire life, riding a bicycle until she was a hundred. It is hard to know what contributed to her longevity, beyond genetics. She smoked for all but the last five years of her life. While this is not an example we should follow, many of us might be tempted to emulate her habit of eating more than two pounds of chocolate every week. While Calment's robust physical condition even late in life was extraordinary, it did not mean that she did not age; for instance, she was blind and deaf for many of her final years.

Calment is the record holder, but one has to remember that she was born almost 150 years ago, in 1875. It is almost a miracle that she survived for so long in the age before antibiotics and other advances in modern medicine. Given the even greater progress made since then, might we expect today's humans to live much longer?

A few years ago, Jan Vijg and his colleagues at the Albert Einstein College of Medicine in the Bronx published a study that analyzed demographic data from several countries to look at shifts in the population of each age group. As life expectancy improves, the fastest growing segment of the population is usually the oldest, since many more people reach the threshold for that group. For example, in France in the 1920s, 85-year-old women were the fastest growing group. By the 1990s, the fastest growing group were 102-year-olds. You might expect that with time, this would shift to even older ages. But the study showed that improvements in survival decline after age 100, and the age of the oldest person has not increased since the 1990s. Vijg predicted that the natural limit of our life span is about 115 years; there will be occasional outliers such as Jeanne Calment, but he calculates that the probability of anyone exceeding 125 in any given year is less than 1 in 10,000.

This conclusion was contradicted a couple of years later by a study examining records of men and women in Italy who had reached the age of 105 between 2009 and 2015. It concluded that mortality rates plateaued after the age of 105, in an apparent violation of Gompertz's law. The researchers went on to say that a limit to longevity, "if any, has not been reached." This paper in turn was criticized by one of the authors of the earlier study, who felt that it was rather far-fetched that after increasing exponentially for most of one's life, the chance of dying should plateau in extreme old age. Others pointed out that most of the cohort did, in fact, follow Gompertz's law, so the plateau came from less than 5 percent of the mortality data. Moreover, they argued that even if mortality did plateau after age 105, the likelihood of anyone surviving much beyond Calment's 122 years was remote, in the absence of major biomedical advances. It is a question of statistics. At today's rates, the odds of surviving each year after 105 is only about 50 percent; to beat Jeanne Calment's 122 would be like tossing a coin seventeen times and having it come up heads every time. Those odds are about 1 in 130,000.

Recent data support the views of Vijg, Olshansky, and other proponents of a limit to maximum life span. After climbing steadily for the last 150 years, the annual increase in life expectancy slowed down globally around 2011 to a fraction of what it had been in previous decades, and plateaued from 2015 to 2019 before falling precipitously as a result of the Covid-19 pandemic. The pandemic, like the influenza epidemic that gripped the world in 1918–19, killing an estimated 50 million people, was an exceptional situation. But we weren't making progress even in the handful of years *before* the pandemic. Why not is unclear. It could be due to the rising epidemic of obesity and associated scourges such as type 2 diabetes and cardiovascular disease. As people live longer, Alzheimer's and other neurodegenerative diseases are responsible for an increasing share of deaths, and there is currently little treatment for them.

In any case, although the number of people who live to be 100 keeps increasing, nobody has beaten Calment's record of 122 in the twenty-five years since she died. The next oldest person, a Japanese woman named Kane Tanaka, died in 2022 at the age of 119. As I write this, the oldest living person is Maria Branyas Morera of Spain, who is 116 years old. What is striking is that these extremely long-lived people are all women. Now that death rates due to child-birth have been reduced dramatically, life expectancy for women is greater than that of men in nearly every country.

Even if nobody beats Calment's record soon, there remains great interest in why some humans live exceptionally long. Thomas Perls, who heads the New England Centenarian Study, has been studying centenarians for several decades. As a practicing physician who specializes in geriatrics, he confronts the realities of aging in his patients every day. He investigates the health history, personal habits, and lifestyles of centenarians, along with what is known about their family histories and genetics. In one large study, Perls concluded that centenarians fell into three classes. About 38 percent were what he called Survivors, who had been diagnosed with at least one age-associated disease before the age of eighty; another 43 percent were Delayers, who developed such a disease after the age of eighty; and the last group consisted of Escapers, the 19 percent who reached their hundredth birthday without being diagnosed with any of the ten most common age-associated diseases. In fact, about half of centenarians celebrated turning one hundred without heart disease, stroke, or non–skin cancer, which is extraordinary.

Perls says that centenarians generally maintain their independence up through their early to mid-nineties. For those who live beyond 105, that independence can be observed at least through age 100. So it appears that centenarians survive for so long by staying healthy longer than most people, rather than going through a prolonged period of living with diseases of old age. Perls also told me that he has seen an increase in the number of people aged 100 to 103, a likely reflection of improvements in medicine and lifestyle over the last few decades, but, beyond that, he is not seeing an increase—perhaps because genetics play such an influential role in survival to those extreme ages. He agrees with Olshansky that currently there is a natural limit on our life span.

Perls and other researchers are now sequencing the genomes of centenarians, and he plans to also study the modifications in DNA that accumulate with age. These studies could reveal the underlying biology of extreme longevity in ways that could be very useful to the rest of us. In the meantime, based on what he has learned so far, Perls has developed a website, livingto100.com, which asks visitors questions about themselves, and spits out an estimated life span, along with suggestions for how to improve it. A few findings may surprise you: it recommends tea over coffee, reducing our intake of iron (often found in multivitamins), and flossing regularly. But many of the suggestions are what one might expect: eating moderately and healthily and avoiding fast food, processed meat, and excessive carbohydrate consumption, as well as exercising and maintaining a healthy weight, getting adequate sleep, reducing stress, staying mentally active, and having an optimistic outlook. It helps not to have diabetes, and having a close family member who lived to be over ninety is a big plus. Since my father, at ninety-seven, still does his own laundry, grocery shopping, and cooking—making complicated Indian recipes and his own ice cream from scratch—I may have lucked out.

The debate about whether there is a limit to human longevity led to a famous bet. At a 2001 meeting, a reporter asked Steven Austad when we would see the first 150-year-old human. None of the other scientists wanted to go out on a limb, but Austad blurted out, "I think that person is already alive." When he read about this, Olshansky, who remains skeptical of exceptional longevity, called up Austad and challenged him to a friendly bet. You might think that this was a safe bet since they would both be dead before it could be decided, but they'd already thought of that. The two men agreed to put $150 each into a fund for 150 years, which, Austad notes, had a nice symmetry to it. A back-of-the-envelope calculation by Olshansky suggested that in 150 years, $150 could turn into about $500 million to be won by either them or one their descendants. A dozen years later, nobody had yet approached the age of Jeanne Calment, but both of them still felt confident, so they doubled the bet, with each putting another $150 into the pot, raising the potential stake to a cool $1 billion 150 years from now—although it is not clear what $1 billion would actually buy at that point.

Why did Austad make this bet? It is not as if he believes that just because we are getting better at treating diseases of old age such as cancer, stroke, and dementia, people will live thirty years more than Calment. In fact, on that point, he and Olshansky agree. Rather, Austad believes that research on aging will result in game-changing medical breakthroughs. The scientists disagree mainly on how rapidly these innovations will occur.

We have now explored how evolutionary theories help us understand why death occurs at all, and how the optimization of fitness by evolution has resulted in a huge range of life spans in different species. We have also explored whether there are biological limits to our own life span. But none of this tells us *how* aging occurs and how it leads to death.

The quest to defeat aging and death is centuries old, but findings from modern biology over the last half century have led to an explosion of knowledge about exactly what goes on in our bodies as we age. As we noted before, aging is simply an accumulation of damage to our molecules, cells, and tissues due to a variety of causes that bring about increasing debilitation and eventually death. An aging body changes in so many ways that it is hard to glean which factors cause aging and which are simply its consequences. But scientists have homed in on a small number of hallmarks of aging. According to them, such a hallmark should have three characteristics: first, it should be present in an aging body. Second, an increased presence of the hallmark should accelerate aging. Third, reducing or eliminating the hallmark should slow aging.

These hallmarks exist at every level of complexity, from molecules, to cells, to tissues, to the interconnected system we call our body. No hallmark exists in isolation; they all influence one another. Thus aging doesn't have one or even a few independent causes. It is a highly intricate and interconnected process.

It is easiest to make sense of it all if we start at the most basic level of complexity: with the molecule that could be thought of as the ultimate command and control center of the cell.

3.

DESTROYING THE
MASTER CONTROLLER

The ancient site of Hampi in South India offers a stark contrast to the thriving metropolis of London. The grand city that existed for more than a thousand years and at its peak in the early sixteenth century was second in wealth only to Beijing is now a collection of well-preserved granite ruins about fifteen miles from the nearest railway station. The once-bustling marketplaces and intricately carved temples and palaces are now only alive with camera-toting tourists. It was once the London of its time: the seat of an empire and a flourishing center of trade and culture. When I travel to London, I simply cannot imagine the city ever not existing, and the inhabitants of Hampi probably thought the same. This failure of imagination extends to us as individuals too. Even if we know we are going to age and die, in our daily lives, unless we are terminally ill, we carry on as if we are immortal.

How could a thriving, vibrant city like Hampi have disintegrated and no longer exist? Throughout history, one of the fastest ways for a society to crumble was the breakdown of law and order resulting from a government's loss of control due to civil unrest or a war. And just as with society, loss of control and regulation in biology leads to decay and death, not only of the cell but of the entire organism.

Unlike a functioning society run by a government, there is no central authority in the cell that supervises its thousands of components as they go about their business. So is there even a counterpart in the cell of a command and control center? Perhaps the closest thing is our genes, which reside in our DNA. The nature of genetic information in our DNA and the ways it becomes corrupted over time are essential for understanding aging and death.

We didn't even know about genes as an entity until the late nineteenth century. Most of us think of genes as traits that we inherit from our parents and pass on to our children. We may think of good genes, reflected in positive traits, or bad ones, characterized by disease or defects. But genes are better described as units of information. They contain information not only on how to reproduce an organism and pass on its traits, but also on how to build an entire organism from a single cell and keep it functioning.

Among the most important information that genes contain is how to make proteins. We normally think of proteins as essential components of our diet, and we know they are used to build muscle. In fact, our body contains thousands of proteins. Not only do they give the body form and strength, but they also carry out most of the chemical reactions that are essential for life. They regulate the flow of molecules in and out of cells. They allow our cells (and us) to communicate with one another. They are the reason we can sense light, smell, touch, and heat. Our nervous system depends on proteins to transmit nerve signals and even to store memory. The antibodies we use to fight infections are proteins. Proteins also enable the cell to manufacture all the other molecules it needs, including fat and carbohydrates, vitamins, and hormones, and—to complete the circle—even our genes. Proteins are everywhere. And every one of these proteins is made by following instructions in a gene.

Exactly *how* genetic information is stored and used remained a huge mystery until relatively recently. Even in the 1940s, scientists still didn't understand the molecular nature of genes. Today we know that our genes reside in DNA, a long molecule that consists of two strands wrapped around each other in a double helix. Each strand of DNA has a backbone made up of alternating groups of phosphate and a sugar called deoxyribose. If that were all DNA was, it would just be like any other repeating polymer such as polyethylene or other plastics, and incapable of carrying information. But DNA is able to encode instructions because each sugar in its backbone is attached to one of four types of chemical groups called bases. These bases are adenine (A), guanine (G), thymine (T), and cytosine (C). This phosphate-sugar-base unit is the building block of DNA, known as a nucleotide.

You can think of each building block as a letter, and a DNA chain as a very long sentence written using this four-letter alphabet. Just as a particular sequence of letters can form a sentence that conveys meaning and information, suddenly you could imagine how DNA could too, but it was still not at all clear how. This changed dramatically in 1953 when the three-dimensional structure of DNA was deduced by James Watson and Francis Crick. Normally, the structure of a molecule only hints at how it might work, but DNA was different. Its structure immediately shed light on how the sequence of bases could transmit information, transformed our understanding of genetics, and ushered in the current revolution in molecular biology. Without it, we would have had no hope of understanding the workings of life or unlocking the secrets of why we age.

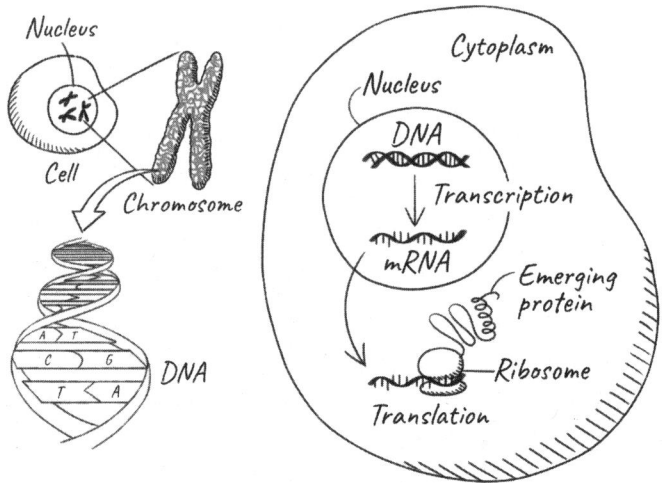

Genetic information stored in our chromosomes in the form of DNA is copied (transcribed) into mRNA in the nucleus. The mRNA then moves to the cytoplasm, where ribosomes read it to make proteins.

In DNA, two strands running in opposite directions are wrapped around each other in a double helix. A base from one strand chemically bonds, or pairs, with the base directly across from it in a very specific way: an A pairs only with a T or vice-versa, and a C with a G. Hence the magic of DNA: if you know the sequence of bases in one of the two strands, you can determine the sequence of the other. This also means that if you separate the two strands, each of them has the information to make the other, enabling you to create two identical copies of the molecule from an original. Suddenly an age-old problem was solved: How could you get two daughter cells, each of them possessing exactly the same genetic information as the single parent cell? Genetics had become chemistry: we could understand at the molecular level how genetic information could be duplicated and passed on to a new generation.

Still, there remained the second question of how genetic

information in DNA actually codes for proteins. It turns out that the section of DNA that codes for a gene is copied into an intermediate molecule called ribonucleic acid. RNA is similar to DNA but with some important differences. Unlike DNA, it has only one strand, and instead of deoxyribose, it has a sugar called ribose. In RNA, the thymine (T) base is replaced by uracil (U), which is slightly different chemically but pairs with A just as T does.

Think of DNA as the collection of all our genes, much as the British Library or the US Library of Congress are collections of all the books published in their respective countries. Those libraries are not likely to let you take a valuable eighteenth-century book home to read at your leisure. But they can often provide a copy of it to take home. Similarly, RNA is a working copy of the gene that can be used by the cell.

Not every piece of DNA that is copied to RNA codes for a protein. Some RNAs are part of the machinery that is used to make proteins. Others can even control whether certain genes are turned on or off. But when an RNA is made from a gene that codes for a protein, it is called messenger RNA, or mRNA, because it carries the genetic message for how to make that protein. We've heard a lot about mRNA recently in connection with vaccines for Covid-19. These vaccines are made from mRNA molecules that contain instructions on how to make the spike proteins that are on the surface of the virus that causes Covid-19. When those mRNA molecules are injected into us, our cells read the instructions in it and produce the corresponding spike proteins, which in turn trains our immune system to be ready to fight the real Covid-19 virus.

How instructions in mRNA are read to make proteins was a hard puzzle that took over a decade to crack. The problem scientists faced was that proteins too are long chains, but of completely different types of building blocks called amino acids. Unlike DNA and RNA, which have four types of bases, there are at least twenty different types of amino acids. If proteins were like sentences written in a twenty-letter alphabet, how could they translate those sentences from the four-letter language of genes? The way nature has solved this problem is that groups of three bases (or letters) in mRNA are read as a code word, or codon, each of which specifies an amino acid. The whole process takes place on the ribosome, a giant, ancient molecular machine that consists of almost half a million atoms.

I have spent much of my life trying to understand how the ribosome carries out the complicated process of reading mRNA to synthesize a protein. What seems miraculous is that as the newly made protein chain emerges from the ribosome, the sequence of its amino acids contains within itself the information needed for the protein chain to fold up into a particular shape so that it can carry out its function. It is akin to writing different sentences on strips of paper and, depending on what I had written, each strip would magically fold itself into its unique shape. This ability of a protein chain to fold itself up is why the one-dimensional information contained in our genes allows us to build the complex three-dimensional structures that make up a cell—and, eventually, us.

The gene doesn't just contain information on how to *make* a protein. The part that specifies that is called the coding sequence, but flanking it are regions (non-coding sequences) that signal *when* to make the protein, when to stop, and even whether to make it quickly or slowly, for a brief while or for a long time. These signals are turned on or off either by chemicals in the environment or by other genes. Genes, in other words, don't act alone; they form a giant network with lots of other genes, as well as the broader environment. This is why some proteins are made by all our cells, but others only by specific cells, such as skin cells or neurons. And why some proteins are made only at certain stages in our development from a single cell to a complete human being. The precise orchestration of this network of thousands of genes is what makes life possible.

You could think of the process of life as an enormous program that somehow activates itself using the blueprint provided by DNA. The word *blueprint* is a convenient metaphor, but we should not take it *too* literally, because a blueprint implies a rigid manufacturing process that produces a strictly defined product. Unquestionably, DNA is the central hub for regulating the overall program of the cell. But I think of the cell as more like a democracy than a dictatorship. Just as an ideal government is not autocratic but responsive to the needs of its people over time, DNA does not dictate the entire process. Rather, conditions in the cell and its environment decide which parts of the DNA are used, as well as how often and when.

UNDERSTANDING THE MOLECULAR BASIS OF genetics has transformed modern biology, but what does it have to do with aging? If the genes in our DNA specify the program of the cell, why doesn't the program just keep running forever? The problem is that the DNA itself changes and deteriorates with time.

Of course, genes and mutations were studied long before we knew about DNA. Prior to DNA, the only way to determine whether an organism had a genetic mutation was when it resulted in a change in an observable trait. Today we know that mutations are simply changes in the bases of DNA. Changing bases in DNA is the equivalent of changing letters in a sentence. Sumtymes we can still dicifer the same meening, but other times, just a single change can be confusing or even have the opposite meaning—for example, if we change the word *hire* to *fire*.

Now that we can sequence DNA—or determine the precise order of bases in any piece of DNA—we can see that mutations happen all the time. Many of them have no observable effect.

This is because even with the change to the DNA, the altered gene functions just as well; or the organism has redundant genes, so that if one is defective, the others can compensate for it. Other mutations can be harmful to varying degrees because they result in proteins that are defective; or proteins that are produced in the wrong amounts or at the wrong time.

Sometimes, mutations can actually be beneficial. For instance, if the mutation occurs in a germ-line cell, it might very occasionally give offspring an advantage that facilitates their survival. A species that is uniformly the same could be wiped out by some pestilence, like trees susceptible to Dutch elm disease, or by sudden changes in the climate or geography. Mutations can give rise to genetic variability in a population and make it more resilient by increasing the likelihood that some strains might survive better than others as conditions change. Without mutations, there would be no evolution; we would never have emerged from primitive molecules. The cell, then, must strike a balance, tolerating enough mutations in the germ line to allow variability and evolution, but not allowing so many mutations in our somatic cells that the complex process of life begins to break down.

A societal breakdown of law and order can bring about chaos, mass starvation—even the annihilation of entire cities and civilizations. The worst criminal elements often take advantage during turbulent times, usurping power and making life miserable for everyone else. Similarly, loss of control in biology can lead to deterioration and death as well as to many diseases. One of the worst examples of cells misbehaving is cancer, in which aberrant cells are no longer inhibited by neighboring cells but instead multiply unchecked and take over entire tissues and organs, interfering with their functioning. In that sense, cancer and aging are intimately related: they both arise from a biological loss of control, and their ultimate source is often mutations in our genes, owing to changes in our DNA.

LONG BEFORE WE KNEW OF DNA, there were hints that environmental agents could cause what we now know to be genetic mutations. As early as the eighteenth century, the English surgeon Percival Pott discovered that the country's chimney sweeps, many of them children, had abnormally high rates of cancer of the scrotum. He attributed this to their excessive, prolonged exposure to the soot and tar from burned coal. In 1915, Yamagiwa Katsusaburo, a professor of pathology at the Tokyo Imperial University, demonstrated that applying coal tar to the ears of rabbits caused skin cancer. These products of coal would later be identified as cancer-causing agents, or carcinogens, but when Pott made his observations, nobody had any idea what cancer was, and even when Katsusaburo reported his results, the link between cancer and genetic mutations was still decades away.

The first direct evidence linking an environmental agent to mutations was discovered by a scientist with a remarkably peripatetic life. Hermann Muller was a third-generation American who grew up in New York City and entered Columbia College (now Columbia University) at the precocious age of sixteen, graduating in 1910. He stayed on at Columbia for his PhD, working with the famous geneticist Thomas Morgan, who had used fruit flies to show that genes resided in the chromosomes in our cells.

Later, Muller moved to the University of Texas, where, in a key experiment in 1926, he subjected fruit flies to increasing doses of X-rays. As he ratcheted up the dose, the number of lethal mutations rose dramatically. Even a modest application of X-rays produced 35,000 times as many mutations than would have occurred spontaneously. Muller's work advanced genetics tremendously by making it much easier to produce mutations, and also raised awareness of the danger of X-rays and other radiation. At the time, people used X-rays rather cavalierly—it was common for shoe sellers to X-ray the feet of their customers in the shoes they were considering.

Like many geneticists in the early twentieth century, Muller was a proponent of eugenics for much of his life and thought of it as a way for improving the human species. Oddly for a eugenicist, he was also quite left wing, a result of his disillusionment with capitalism in the wake of the Great Depression. He recruited lab members from the Soviet Union and as a faculty advisor, helped edit and distribute a leftist student newspaper called *The Spark*, which spurred the FBI to investigate him.

Partly as a result, in 1932 Muller left the United States for Berlin. Discouraged by the rise of Hitlerism, he left the following year for the Soviet Union, believing that the environment there would be more conducive to his left-wing views. He spent a year in Leningrad before moving to Moscow for a few years. He had not, however, reckoned with the rise of Trofim Lysenko, the Soviet biologist and charlatan who had ingratiated himself to Stalin. Lysenko viewed genetics as inconsistent with socialism, and instead espoused a number of crazy ideas in agriculture, while ruthlessly wielding his power to suppress or destroy any biologist who dared question him. In doing so, he contributed to famines that killed millions of people and set back Soviet biology by decades. Muller and other geneticists did what they could to counteract Lysenko, but eventually Muller incurred Stalin's wrath for his views on both genetics and eugenics and had to flee.

Not yet ready to return to the United States, where the FBI was still investigating him, Muller ended up at the Institute for Animal Genetics at the University of Edinburgh in 1937. There he helped catalyze another important discovery. He joined a lively group of scientists, many of them refugees from totalitarian regimes, under the direction of pioneering medical geneticist Francis Crew.

One of Crew's key collaborators, Charlotte Auerbach, had been born to an academic Jewish family in Krefeld, Germany. Auerbach, known as Lotte, was an independent thinker who did not take well to being told what to do. While studying for her PhD in Berlin, her professor refused her request to change her project, so she simply quit and became a high school teacher. She found teaching and keeping order in class exhausting, perhaps not helped by the increasing anti-semitism of the time. In what turned out to be a blessing in disguise, she was summarily dismissed in 1933 at the age of thirty-four because she was Jewish. On her mother's advice, she left Germany, and, with the help of friends of the family, was able to finish her PhD at the Institute for Animal Genetics, where she worked with Crew. In 1939 she became a British citizen; later that year, her mother showed up in Edinburgh without any money or baggage, having made it out of Germany just two weeks before World War II broke out.

Crew's initial attempt to bring Auerbach and Muller together was not a success. He introduced her to Muller and simply told him, "This is Lotte, and she is going to do cytology for you." But Auerbach had no interest in spending her time peering through a microscope to characterize Muller's cells, and, independent minded as always, she refused. She told Muller that she was really interested in how genes enabled development. To his credit, Muller told her that he wouldn't dream of having someone work with him on a project that didn't interest her. However, he persuaded Auerbach that if she wanted to pursue her interest in understanding the role of genes in develop-ment, she needed to produce mutations in them and see their effects.

Around this time, a colleague of hers, Alfred J. Clark, had no-ticed that soldiers exposed to mustard gas in World War I exhibited lesions and ulcers that resembled the effects of exposure to X-rays. Auerbach, along with Clark and their colleagues, exposed fruit flies to mustard gas, checking for mutations using the methods Muller had pioneered. It says something about their dedication that their experiments were carried out on the roof of the Pharmacology Department in cold, wet, blustery Edinburgh. The experimental conditions would never pass a workplace health and safety inspection today: the fruit flies were exposed to the gas in vials and afterward were removed by hand, causing serious burns to the workers. In any case, the results were unambiguous. Exposure to mustard gas had resulted in ten times as many lethal genetic mutations. Chemicals, like radiation, could also cause mutations.

MULLER AND AUERBACH'S WORK SHOWED how our genetic blueprint could be damaged by environmental agents such as radiation or chemicals. At the time, we didn't even know that DNA was the genetic material, let alone how the information it carried could be corrupted. But once Watson and Crick revealed its double-helical nature, the question naturally became how exactly did these agents cause changes in our DNA that resulted in mutations?

Studying the biological effects of radiation had been something of a stepchild of the life sciences before World War II. But once the world saw the horrible effects of radiation wrought by the two atomic bombs dropped on Japan in August 1945, the US government became very interested in this once sleepy field. After the war, many of the sites that had been used for the Manhattan Project to develop nuclear weapons were converted to radiation biology research centers. One of these was Tennessee's Oak Ridge National Laboratory, which had originally been the site for producing large amounts of the uranium isotope used in the first atomic bomb, detonated over the city of Hiroshima. Remote from the large academic centers of the United States in the Northeast and the West Coast, Oak Ridge was nestled between the spectacular wilderness of the Cumberland and Smoky Mountains. These attractions, and the generous funding provided by the government, allowed Alexander Hollaender, a leading radiation biologist of his time, to recruit many excellent scientists to Oak Ridge, including Dick and Jane Setlow.

Dick and Jane Setlow met as undergraduates at Swarthmore College in the 1940s and married soon afterward. When Hollaender approached them around 1960, Dick was on the biophysics faculty at Yale University. It was one of the oldest biophysics programs in the country, but Hollaender lured away Dick with a shrewd move: he offered Jane, who had a temporary appointment working for someone else, a full position too. In those days, even women who had earned graduate degrees rarely had the opportunity to work as equals and ended up assisting some male scientist, frequently their husband. Hollaender's gambit worked. Both Dick and Jane became leaders in the field, sometimes working together but just as often separately. They also raised a family of four children and hiked and hunted for fossils in the mountains around Oak Ridge before moving to another national lab in Brookhaven on Long Island about fifteen years later.

Brookhaven National Laboratory was where I first met them, in 1982. Dick was the chair of the department that hired me. It might have helped that I was desperately trying to leave Oak Ridge after only fifteen months there because the resources I had been promised never materialized. Dick, having made the same move himself, was sympathetic. At the time, I was thirty-one years old, and although they were only around sixty then, I regarded them as ancient fossils, like the ones they collected. Like some of the more mainstream molecular biologists, I severely underestimated the importance of their work, and I regret that I didn't talk to them about their discoveries when I had the chance. It's a reminder to me of how insular most scientists are, with little appreciation of what goes on outside their narrow specialties.

Even before X-rays were discovered, we knew about other forms of radiation. As early as 1877, the British scientists Arthur Downes and Thomas Blunt discovered that sunlight could kill bacteria. In the early twentieth century, Frederick Gates showed that it was the shorter wavelengths in sunlight—ultraviolet, or UV, radiation—that had the killing effect. Soon after Muller demonstrated that X-rays could cause genetic mutations, scientists started studying UV radiation too; after all, it was easier to produce and safer to handle. They found that for a given dose, UV light produced even more mutations. At Oak Ridge, Dick and Jane began by trying to understand exactly how UV caused mutations in DNA. One finding that intrigued them was that UV light links up two adjacent thymines (the T bases) on DNA. Virtually any sequence of DNA will occasionally have two thymines next to each other, and somehow UV was linking them together so that the two bases were no longer separate but acted as a single unit consisting of two building blocks—known as a thymine dimer, or sometimes as a thymidine dimer, if scientists want to refer to the larger unit that includes the sugar to which the thymine is attached. Was this how UV inactivated DNA and killed bacteria?

Dick and Jane experimented with inserting foreign DNA into a bacterium. This enabled them to introduce a gene that gave the bacterium new abilities, such as growing in the absence of a nutrient it would need otherwise or becoming resistant to an antibiotic. However, when they tried this using DNA containing thymine dimers, it was as if the DNA had become inactivated. Dick went on to show that thymine dimers prevent the DNA from being copied, so new DNA could not be made.

The next step was even more remarkable. Dick and his colleagues found that shortly after exposure to UV radiation, the thymine dimers disappeared from the DNA altogether. The dimers, including the sugar and phosphate to which the bases were attached, were cut out of the DNA, with the missing section filled in using the other strand as a guide, just as when DNA is copied. Discoveries in science are not made in a vacuum. The state of knowledge reaches a stage where the next advances are possible, so new breakthroughs are often made simultaneously. The same year, 1964, that Setlow reported his discovery, two other groups, led by Paul Howard-Flanders and Philip Hanawalt, respectively, made similar findings. The reports all confirmed that the cell clearly had some mechanism to not only recognize the thymine dimers but also to repair them, by a process called excision repair.

Excision repair was also found in a different context. Even in the 1940s, scientists realized that they could reverse the effects of UV light on bacteria by exposing them to visible light. The arrested bacteria would start growing again. Extracts from bacteria that had been exposed to visible light could repair damaged DNA. How it worked was something of a mystery until Aziz Sancar, a Turkish doctor turned scientist, got involved in the work and identified its mechanism, which also involved repairing thymine dimers using a different enzyme. Oddly, *Hemophilus influenzae*, the organism in which Dick Setlow had identified the same kind of repair, lacks this mechanism (as do we humans)—otherwise he might never have made his discovery. Just the fact that nature had evolved two completely different mechanisms to remove thymidine dimers tells us about the importance of repairing them.

These experiments established firmly that the cell could repair damaged DNA. But we're rarely exposed to high doses of X-rays. Our clothes and the melanin pigment in our skin protect us from a lot of UV exposure. Also, we know enough to stay away from mustard gas, coal tar, and other nasty chemicals, which human beings never encountered in the wild in prehistoric times. Yet these mechanisms to repair damaged DNA evolved billions of years ago and are part of every life form.

It turns out that our DNA is constantly being assaulted, even in the normal course of living, without exposure to nasty chemicals or radiation. The person who did more than anyone to make us appreciate this was the Swedish scientist Tomas Lindahl. As a postdoctoral fellow at Princeton University, he was working on a relatively small RNA molecule. To his frustration, he found that it kept breaking down.

As we've discussed, RNA molecules use the sugar ribose rather than the deoxyribose found in DNA. Ribose differs from deoxyribose by just one additional oxygen atom. That extra atom makes RNA much more unstable, but also gives it the ability to form complex three-dimensional structures that can carry out chemical reactions. Because of these properties, scientists believe that life originally emerged in a primordial world in which RNA carried out chemical reactions as well as stored genetic information. As life evolved to become more complex, using an unstable molecule to store an increasingly large genome was not viable, and so the more stable DNA was used to store genetic information.

Lindahl knew that DNA was more stable than RNA, but he wanted to know how much more. It had to be stable enough to pass on information to the next generation without too much change. Or over the billions of cell divisions that occur by the time a single cell develops into a mature organism. That is a *very* long time.

Lindahl studied DNA in a variety of conditions and found that over time some of its bases changed. The most common change was that the base cytosine (C) was transformed into a different base called uracil (U), which is normally found in RNA, where it stands in for thymine (T). The problem is that, like T, U pairs with an A, while C pairs with a G. This transformation was like changing a letter in the DNA sentence. Having many of these changes throughout the genome would corrupt the encoded instructions to the point where they would become nonsensical.

Lindahl showed that the change from a C to a U can be caused simply by exposure to water, a ubiquitous occurrence for all living molecules in a cell. In one day, water could cause about ten thousand changes to the DNA in each of our cells. Lindahl estimated later that, taking into account all forms of spontaneous damage to DNA, about a hundred thousand changes are inflicted on the DNA in each of our cells every single day. It was hard to imagine how life could survive when the set of instructions that enabled it was being corrupted so rapidly. Clearly, there had to be a mechanism to correct these errors too. Over the next few decades, Lindahl and other scientists worked out how this change is repaired.

A much more drastic form of DNA damage occurs when both strands break, leaving two pieces that have to be rejoined. Sometimes there are even multiple breaks on different chromosomes. This can result in a complete mess, where half of one chromosome is joined to the other half of a completely different one, or where a broken-off piece has been reinserted backward. Again, if we think of DNA as a text consisting of sentences, changes to individual bases are like typos: although they will occasionally garble the meaning, often you can still make sense of them. But if you repair a double-strand break incorrectly, it is like cutting sentences or whole paragraphs from a long text and pasting them back in some random order. Occasionally, it might still sort of make sense, but other times it will be complete gibberish. So it is imperative for the cell to join broken ends of DNA as soon as it recognizes them, preferably before multiple breaks occur. Special proteins recognize the broken ends and join them together to make an intact DNA molecule. This process does take into account the DNA sequence at the ends, so if there is more than one break in the cell at any given time, there is always a chance that it will join the wrong ends. When our genome is scrambled in this way, it can lead to different kinds of problems. One is a loss of function, where the cell cannot do its job efficiently or perhaps not at all. In other cases, it can corrupt or lose the signals that control genes. As a result, the cell starts growing unchecked, leading to cancer.

Humans are what we call diploid, possessing two copies of each chromosome. The more common and accurate way that the body repairs double-stranded breaks is to use the undamaged DNA in the other chromosome as a guide. Even in organisms such as bacteria, a second copy is often present when cells are dividing and the DNA is being duplicated. Either way, the repair machinery lines up the broken ends against the matching sequence on the other (intact) copy of the DNA to form a complicated structure in which all four strands are intertwined. This is more accurate than simply grabbing random ends and joining them because it checks whether they are the right ends to be joined. By doing so, it restores the integrity of the genome and fills in any gaps that arise if the broken ends have been frayed.

Apart from chemical damage, mutations have another way of creeping into our genome. Each time a cell divides, the entire genome has to be duplicated, which is like copying a text three billion letters long. No process in biology is ever completely accurate. Just as with writing or typing, the faster you try to copy something, the more prone you are to making mistakes. The polymerase enzymes that replicate DNA are incredibly accurate; what's more, they can proofread their work, so to speak, correcting mistakes as they go. Nevertheless, they still make an error once every million or so letters. In a genome with a few *billion* letters, that means several thousand mistakes occur each time the cell divides. The cell can't take forever to divide, and in life there is always a compromise between speed and accuracy. Not surprisingly, the cell has evolved sophisticated machinery to correct these errors.

Relying on some very clever experiments, Paul Modrich figured out how enzymes in a bacterium recognize the mismatch, cut out a section of the new strand containing the mistake, and fill in the section so that the mistake is corrected. That mechanism is now well established in bacteria, but scientists are still debating exactly how these kinds of errors are corrected in higher organisms like humans.

It took a long time for the scientific community to realize the importance of DNA damage and repair. Muller received the Nobel Prize in 1946, a full twenty years after his discovery that X-rays cause mutations. But by the time the 2015 Nobel Prize in Chemistry went to Lindahl, Sancar, and Modrich, the field of DNA repair had long ceased to be a scientific backwater. Now it is widely recognized as crucial for life as well as for understanding the basis of both cancer and aging. As in most scientific areas, hundreds of scientists working in different labs throughout the world had contributed to these discoveries, but the Nobel Prize can be shared by only three people at most, so the committee has the unenviable job of choosing the three most important to honor, not always without controversy. The prize also cannot be given posthumously, and, sadly, Dick Setlow had died a few months before it was announced, at the age of ninety-four.

Over the years, scientists have isolated many different repair enzymes. Many of them are essentially the same in all life forms from bacteria to humans. DNA repair is so essential to life that it originated billions of years ago, before bacteria and higher organisms diverged. Maintaining the stability of the genome and its instructions is critical for the cell and demands constant surveillance and repair. You can think of these repair enzymes as the sentinels of our genome.

Because DNA damage occurs all the time, any defect in the repair machinery itself is particularly disastrous because it means that the damage would accumulate rapidly. Not surprisingly, many mutations in the repair machinery have been linked to cancers: for example, mutations in the BRCA1 gene predispose women primarily to cancers of the breast and ovary. Defects in the repair machinery also cause aging, but because we are also more likely to develop cancer as we age, it is hard to separate out the two effects. Perhaps more than any single person, the Dutch scientist Jan Hoeijmakers has worked extensively to explore how DNA repair defects can age a person prematurely. One condition he has focused on is Cockayne syndrome, which manifests symptoms associated with aging, such as neurodegeneration, atherosclerosis, and osteoporosis. In females, defects in how the cell responds to DNA damage can affect the age at which menopause begins. Generally, the more effectively our bodies can repair our DNA, the more we can resist aging.

WHEN A CELL SENSES SIGNIFICANT DNA damage, it triggers what is called the DNA damage response. This is not *all* good news: the damage response often has greater consequences for aging than the damage itself. Sometimes the cell will go into senescence, a state in which it is unable to divide further, and in extreme cases, the cell is triggered to commit suicide. It is odd to think that life would have evolved a mechanism to kill its own cells, but one individual cell among an organism's billions is ultimately dispensable. If, however, that cell were allowed to become cancerous as a result of DNA damage, it could multiply and eventually kill the entire organism. Both cell death and senescent cells are important factors in aging, especially the latter, and we will have a lot more to say about them in later chapters. Suffice it to say here that the DNA damage response evolved to balance the risk between cancer and aging. It is one more mechanism that evolved to benefit us early in life, even if it costs us later, after we've already passed on our genes.

At the heart of the damage response is a protein called p53, the product of the TP53 tumor suppressor gene. This protein is so essential that it is often called the Guardian of the Genome. Almost 50 percent of all cancers have a mutation in p53; in some forms of cancer, the rate is as high as 70 percent. Normally, p53 is bound to a partner protein and is inactive. It is also turned over rapidly in the cell, so it is made and then degraded all the time. When DNA damage is sensed, p53 is activated and starts to accumulate. It is also freed from its partner protein, springs into action, and turns on the expression of many genes; in this context, expression means the production of the functional protein from the information coded by the genes. Some of them are genes for DNA repair proteins. Others stop the cell from dividing to give DNA repair genes a chance to do their job. When the damage is too extensive, p53 can turn on genes that induce cell death.

P53 may also hold the key to Peto's paradox, an oddity observed in the 1970s by the British epidemiologist Richard Peto. Large animals such as elephants or whales can have a hundred times as many cells as we do. Even accounting for their slower metabolism, this means there is a much greater chance that one of their cells will mutate to become cancerous. Yet these large mammals are remarkably resistant to cancer and live almost as long or even longer than us. Humans inherit one copy of the gene for p53 from each of our parents, but it turns out that elephants have *twenty* copies. Therefore their cells are exquisitely sensitive to DNA damage and commit suicide when it is detected. Scientists are always worried about proving cause, so they wanted to find out what would happen if you increased the level of repair genes in other organisms. Curiously, in studies involving fruit flies, they found that repair gene overexpression did indeed increase longevity—but only if the genes were turned on throughout the fly's entire life. If the repair genes weren't activated until adulthood, there was no increase in life span.

Some of the long-lived species we encountered in chapter 2, such as certain whales and giant tortoises, also have unusual variations in the numbers and types of tumor suppressor genes. Perhaps without this, they would have died of cancer at much younger ages. In general, there seems to be a powerful correlation between strong DNA repair genes and longevity. Humans and naked mole rats, which can live up to 120 and 30 years, respectively, have a higher expression of DNA repair genes and their pathways than do mice, which live only up to 3 or 4 years. It remains to be seen whether exceptionally long-lived people have unusually efficient DNA repair mechanisms.

Paradoxically, many new cancer therapies work by inhibiting DNA repair. This is because cancer cells have defects in some of their repair machinery, so inhibiting other routes of repair closes off their options. Unable to repair their own DNA, the cancer cells die off. However, this is a short-term solution to combating aggressive cancers; normally, blocking DNA repair over an extended period could actually increase a person's risk of both cancer and aging. Attempting to use our knowledge of DNA damage and repair to tackle aging is not straightforward because of the tricky interplay between aging and cancer.

Even if it is difficult to use DNA repair to directly improve longevity, our knowledge of it underpins our understanding of virtually every process of aging. Genes ultimately control the entire process of life: when and how much of each protein we make; whether our cells continue to live or suddenly stop dividing; how well our cells sense nutrients in their surroundings and respond to them; and how different molecules and cells communicate with one another. Genes control our immune system, which must maintain the delicate balance of reacting to invading pathogens without inducing chronic inflammation.

Direct damage to our DNA, and the cell's seemingly paradoxical response to it, is only one of the ways our genetic program can be changed as to cause aging. For our DNA has two peculiarities. The first is that its end segments are special and protected, and the consequences of disrupting them are serious. The second is that the way our genome is used does not depend exclusively on the sequence of bases in the DNA itself. Our DNA exists as a tight complex with ancient proteins called histones, and both the DNA and its partner proteins can be altered by our environment to affect the way our genes are used. Our genome, it turns out, is not written in stone but can be modified on the fly.

4.

THE PROBLEM

WITH ENDS

Over a century ago, a scientist in a New York laboratory peered at the cells he had cultivated in flasks and wondered whether he might have uncovered the secret of immortality.

Alexis Carrel was a French surgeon who by then was already famous for having pioneered techniques to reconnect blood vessels that had been severed in an accident or an act of violence such as a stabbing. His method for joining blood vessels end to end with tiny, almost invisible sutures transformed many kinds of surgery, and is the basis of organ transplants even today. In 1904 Carrel left France for Montreal and then Chicago. Two years later, he moved to New York City to become one of the earliest investigators at the newly created Rockefeller Institute for Medical Research (now Rockefeller University). The institute offered an unparalleled environment for an ambitious scientist, including superb laboratories and sizable endowments. And the thirty-three-year-old Carrel certainly had ambitions.

As a surgeon, Carrel dreamed of keeping tissues alive outside the human body. In the lab, we can grow cultures of bacteria or yeast indefinitely. Although individual bacteria or yeast can age and die, the culture continues to grow and is, in a sense, immortal. But that was not clear for cells and tissues from higher life forms such as us. At Rockefeller, Carrel began a long series of experiments to see whether a culture of cells from a tissue could be kept alive indefinitely. By placing the cells from the heart of a chicken embryo in a special flask, and steadily supplying them with nutrients, Carrel seemed to have made a breakthrough. The culture could be maintained for years. These cells, he claimed, were immortal.

The discovery was reported with great fanfare. If cells from a tissue could be made immortal, journalists reasoned, then so could entire tissues and eventually us. An editorial in the July 1921 issue of *Scientific American* gushed, "Perhaps the day is not far away when most of us may reasonably anticipate a hundred years of life. And if a hundred, why not a thousand?"

But Carrel was wrong.

Initially, his work went unchallenged because of his stature, and, over the years, the immortality of cultured cells became dogma. That is, until three decades later, when a young scientist at the Wistar Institute in Philadelphia, Leonard Hayflick, wanted to see if cells would change when exposed to extracts from cancer cells. He decided to use Carrel's method to grow human embryonic cells in culture. To his disappointment, he found he could not grow these cells indefinitely. Initially, Hayflick, a recent PhD in medical microbiology and chemistry, thought he must have made a mistake. Perhaps he hadn't correctly prepared the nutrient broth or was washing his glassware improperly. But over the next three years, he carefully ruled out any technical problems and concluded that the prevailing theory was simply incorrect: normal human cells would not replicate indefinitely in culture. They were not immortal.

Instead, Hayflick found that his cells would divide a finite number of times and then stop. In an ingenious experiment, he and his colleague Paul Moorhead took male cells that had already divided many times and mixed them with female cells that had divided only a few times. When they soon reached their limit, the male cells stopped dividing, while the female ones continued to grow to the point that they came to dominate the culture. Somehow the old cells remembered they were old, even when surrounded by young cells. They were not rejuvenated by the presence of the young cells, nor did they stop dividing because of some contaminating chemicals or viruses in the environment. Hayflick and Moorhead coined the term *senescence* to describe this state, in which the cells were arrested and could no longer divide further.

Another junior scientist might have been nervous about challenging such established ideas, but not the confident Hayflick. He and Moorhead wrote up their results in a meticulously detailed thirty-seven-page paper and submitted it to the same journal in which Carrel had published his original findings. Because it went counter to the prevailing dogma, and perhaps because the editor was a colleague of Carrel's and more inclined to trust him than some young unknown scientist, the paper was rejected but eventually published in *Experimental Cell Research* in 1961. It has since become a classic in the field. The number of times a particular kind of cell can divide is now called the Hayflick limit.

How did Carrel get it so wrong? One possibility, suggested by Hayflick himself, is that the French scientist may have inadvertently introduced fresh cells into the culture each time he replenished the nutrient broth in which they were growing. Some have even suggested that fresh cells may have been incorporated deliberately, although this would be a case of either egregious misconduct or sabotage.

My sneaking suspicion is that by the time Carrel worked on these cells, fame and power had gone to his head, and he had become arrogant and less self-critical about his research. This attitude manifested itself in other ways. In 1935 he published a book titled *Man, the Unknown*, which recommended sterilizing the unfit and gas chambers for criminals and the insane, and commented about the superiority of Nordic people over southern Europeans. In the preface to the book's 1936 German edition, he praised the Nazi government of Adolf Hitler for its new eugenics program. Given Carrel's stature, it is quite possible that the Nazis used his remarks as one justification for their activities. His plaque in Rockefeller University was recently corrected to reflect his views.

Titia de Lange, a renowned biologist currently at the very same Rockefeller University, suggested a more straightforward explanation for Carrel's results: the laboratory next door to Carrel's was working with malignant tumors in domestic chickens, and these cancerous chicken cells might have contaminated Carrel's cultures growing nearby. Cancer cells are the exception to the Hayflick limit: they *don't* stop dividing after a certain number of divisions, and this uncontrolled growth is why cancer wreaks such havoc on the body.

Why don't cancer cells stop growing unlike the normal ones studied by Hayflick? And how can a cell keep count of the number of times it has divided and know when to stop?

When a cell divides, each of the DNA molecules in our chromosomes has to be copied. Unlike bacteria, whose genome consists of a circular piece of DNA, the DNA in each of our forty-six chromosomes is linear. Like an arrow, each strand of the double-helical DNA molecule has a direction, and the two strands of the DNA molecule run in opposite directions. The complex machinery that copies each DNA molecule uses each strand as a guide to make the opposite or complementary strand, but it can do so only in one direction. In the early 1970s James Watson of DNA fame and a Russian molecular biologist named Alexey Olovnikov both noticed at about the same time that the way the cell's machinery copies DNA would create a problem at the very ends of the molecule.

One day, Olovnikov was obsessing over this idea while standing on the platform of a train station in Moscow. He imagined the train in front of him as the DNA polymerase enzyme that copies DNA, and the railway tracks as the DNA to be copied. He realized that the train would be able to copy the rail track ahead of it, but not the part that lay immediately under it. And because the train could go in only one direction, even if it started at the very end of the track, there would always be a section underneath the train that could not be copied. This failure to copy the very end of a DNA strand meant that each newly made strand would be just a little shorter than the original. With each cell division, the chromosomes would progressively shorten, until eventually they lost essential genes and could no longer divide, thereby reaching their Hayflick limit. The end replication problem, as this is known, could explain at least in principle why cells stopped dividing, although the real answer, as we will see, is more complex.

A SEPARATE MYSTERY REMAINED UNANSWERED. Why didn't the cell see the ends of chromosomes as breaks in the DNA and try to join them together? Why didn't it induce some sort of DNA damage response?

In the 1930s and 1940s, around the time that Hermann Muller was investigating how X-rays might damage chromosomes, a young scientist named Barbara McClintock was looking at the genetics of maize. At some point, she discovered the phenomenon of "jumping genes": where genes hop from their position on DNA to a completely different position on the chromosome or even to a completely different chromosome.

Even in the 1930s, both Muller and McClintock, working independently, noticed that there was something special about the ends of chromosomes. Unlike broken chromosome ends, which would often be joined up, the ends of intact chromosomes seemed to stay separate. Muller named the natural ends of chromosomes telomeres. He and McClintock both suggested that they had some special property that prevented them from being mistaken for breaks in the DNA and being joined with each other. This allowed chromosomes to be maintained stably as individual entities in cells instead of being combined randomly. But what made telomeres so special?

Elizabeth Blackburn grew up along with her seven siblings and a large menagerie of pets in the small town of Launceston on the north coast of Tasmania, Australia. She became interested in science and majored in biochemistry at the University of Melbourne, where she had the good fortune to meet Fred Sanger, the famous biochemist who was visiting from England. Encouraged by this encounter, and at a time when there were few women in molecular biology, Blackburn went on to do her doctoral work in Sanger's laboratory in Cambridge. Her timing couldn't have been better, for Sanger had just figured out how to sequence DNA. And there was a second fortuitous event in her life: in Cambridge, she met her future husband, American John Sedat, who soon accepted a position at Yale University. As a result, she decided to join Joseph Gall's lab at Yale for her postdoctoral research.

Gall, a well-established cell biologist, was interested in chromosome structure, and Blackburn knew how to sequence DNA from her work with Sanger. They applied their combined expertise to identify the sequence of DNA specifically at the telomeres of chromosomes. Humans had a mere ninety-two telomeres in each cell; two for each of the forty-six chromosomes. This, they realized, was not enough material. Cleverly, they chose a single-celled organism called *Tetrahymena*, which in one phase of its life cycle has up to ten *thousand* small chromosomes. They found that the sequence of DNA at the telomeres of chromosomes was different not only from anything in the rest of the chromosomes but also from anything they'd ever seen before. TTGGGG (or the complementary CCCCAA on the other strand) was repeated anywhere from twenty to seventy times.

Shortly after Blackburn had characterized these repeats, she encountered Jack Szostak, who was working at Harvard Medical School and was trying to insert artificial chromosomes into yeast. The idea was to introduce new genes into yeast through these artificial chromosomes, which would be replicated along with the yeast's own chromosomes. For some reason, however, they were unstable. The yeast cells were seeing the ends of these artificial DNA molecules as breaks due to damage and setting off a response. Szostak and Blackburn collaborated to see what would happen if they tacked on the telomere sequence of the *Tetrahymena* chromosomes to the ends of Szostak's artificial chromosomes. It worked like a charm: the modified artificial chromosomes were now stable in yeast. Szostak went on to characterize the telomeric DNA from yeast itself. It turned out to have a similar repeat to *Tetrahymena*. Instead of TTGGGG, the repeat was a combination of TG, TGG, or TGGG. From later work, we know now that in humans and other mammals, the repeat is TTAGGG.

Somehow these short telomere sequences told the cell that they were special and should not be treated as ends of broken DNA. Amazingly, although *Tetrahymena* and yeast are separated by more than a billion years of evolution, the slightly different repeat sequence from *Tetrahymena* still works in yeast. This suggests a universal mechanism that protects the telomeres of chromosomes and depends on these repeated sequences.

You could think of these repeated sequences as extra, dispensable material tagged on to the ends of chromosomes. Each time the chromosome replicated, it would lose some repeats, but it wouldn't matter until you eventually lost them all and started losing important genes near the ends of chromosomes. It could explain why cells divided only a certain number of times before they reached the Hayflick limit and stopped.

Even though this explained some things in principle, it still left several basic questions unanswered. What added these telomeric sequences? And why can some cells divide many more times than the Hayflick limit, such as cancer cells or our own germ-line cells?

The first big advance toward answering these questions came when Blackburn, who was now running her own lab at the

University of California, San Francisco, was joined by a graduate student, Carol Greider. The two of them discovered an enzyme that adds the telomeric repeat sequences to the ends of chromosomes. They named it *telomerase*.

Cells from most tissues make very little or no telomerase, but cancer cells and some special cells such as germ-line cells do. Without telomerase, our telomeres get shorter and shorter with age until the cell is triggered into senescence and stops dividing. By contrast, cells with telomerase can simply rebuild their telomeres after each division and thus divide indefinitely. Even introducing telomerase into normal cells can extend their life spans.

As is often the case in biology, it is not *quite* this simple. Cells lose much more DNA during each division than Watson and Olovnikov would have predicted. Moreover, they stop dividing even before all of the telomeric region is lost. And finally, even if telomeres have a special sequence, it still wasn't clear why the cell didn't see them as breaks in the DNA and turn on its DNA damage response.

It turns out that the telomeric ends have a special structure in which one DNA strand extends beyond the other. This longer strand loops back and forms a special structure with the help of special proteins collectively called shelterin, because they shelter and protect the ends of the DNA. This crucial structure is why the cell doesn't recognize the ends of chromosomes as double-strand breaks. A loss or deficiency in shelterin can be lethal, and even moderately defective shelterin can lead to chromosome abnormalities and premature aging, even when the telomeres are of normal length.

When enough of the telomere DNA is lost, these special structures cannot form. The cell then sees the unprotected ends of the DNA as breaks and sets off the damage response, instructing other cells to either commit suicide or go into senescence. We still don't know how or why some cells, like the ones Leonard Hayflick studied, go into senescence while others self-destruct. Perhaps cells that are especially important for maintaining or regenerating tissues—such as stem cells—preferentially commit suicide to avoid passing on damaged DNA to their offspring.

This is all very well for understanding cells in culture, but does this have anything to do with why we age? Or our life spans? And why is telomerase switched off in most of our cells? If we switched it on again, would we simply stop aging?

People with defective telomerase, or who have less than the normal amount of it, prematurely develop a number of diseases associated with old age. Likewise, a stressful life can often make us appear to age faster. We look haggard, and even our hair can turn prematurely gray or white. Stress can also bring on many of the diseases we associate with old age. Stress has multiple effects on our physiology, and exactly how it affects the aging process is complex. But one of the things it does is to accelerate telomere shortening. When we are stressed, our body produces much more cortisol—referred to as the stress hormone—which reduces telomerase activity.

You might expect that species with longer telomeres would live longer, but mice, which typically live only about two years in the lab and much less in the wild, have much longer telomeres than we do. So it may be that the shortening of their telomeres occurs more rapidly. Nevertheless, if you reactivate telomerase in mice that are deficient in the enzyme, you can reverse the tissue degeneration that occurs with aging. According to a number of studies, mice engineered to have even longer telomeres showed fewer symptoms of aging and lived longer. Presumably, starting off with much longer telomeres compensated for their more rapid shortening in mice.

Based on studies like these, many biotech companies are introducing the gene for telomerase into cells or using drugs to activate the telomerase gene that already exists. Some of them are working on how to turn on the enzyme transiently, to avoid the potential problem of triggering cancer by having telomerase switched on permanently. Initially, many of these experiments are focusing on specific diseases where aberrant telomere shortening is thought to be the cause. But the efficacy and long-term consequences of these strategies remain unknown.

When telomerase was discovered, it stirred a lot of excitement in cancer research. Since cancer cells had activated telomerase, scientists thought of it as an anti-cancer target—if you could inhibit it or turn it off, you might kill cancer cells. On the other hand, turning it off could potentially accelerate the shortening of telomeres, which could not only lead to premature aging or other diseases, but by disrupting our telomeres, lead to chromosome rearrangements, which, ironically, could itself cause cancer. There seems to be a delicate balance between telomere loss and aging on the one hand and increased risk of cancer on the other, and it may be that our normal process of switching off telomerase in most of our cells is actually a mechanism to suppress cancer early in life. This balancing act is also apparent from a study showing that people with short telomeres are prone to degenerative diseases, including organ failure, fibrosis, and other symptoms of aging. On the other hand, those with long telomeres face increased risks of melanoma, leukemia, and other cancers. This suggests that we have some way to go before tinkering with telomerase can be a viable strategy for either cancer or aging.

In the last two chapters, we've talked about how genes contain the program to control the complex process of life. In chapter 5, we will see how even allowing for changes from damage to DNA or to our telomeres, the script of life written in our DNA is not fixed. It is modified and adapted on the fly, depending on its history and environment. The ability to annotate the script, much like a conductor would a score or a film director would a screenplay, is the basis of some of the most fundamental processes of life, including how an entire animal develops from a single cell. When the annotation goes awry, that too is a fundamental cause of disease and aging.

5.

RESETTING THE

BIOLOGICAL CLOCK

On June 26, 2000, President Bill Clinton and British prime minister Tony Blair, each flanked by some of the world's most distinguished scientists, linked up via satellite to make a carefully choreographed announcement of "another great Anglo-American partnership." The occasion was the publication of the draft sequence of the entire human genome: the precise order of bases in nearly all of our DNA.

Excitement over this milestone was unanimous across the belief spectrum. Clinton said, "Today we are learning the language in which God created life," while Richard Dawkins, the evolutionary biologist and passionate atheist, said, "Along with Bach's music, Shakespeare's sonnets, and the Apollo space program, the Human Genome Project is one of those achievements of the human spirit that makes me proud to be human."

Other scientists and the popular press gushed with similarly hyperbolic statements. The identification of every human gene would make possible new treatments against diseases and usher in a new era of truly personalized medicine. If we sequenced the genes of individuals, some suggested, we would be able to understand their fate in detail: their strengths and weaknesses, aptitudes and talents, susceptibility to disease, how quickly they would age, and how long they would survive.

The announcement ceremony was the culmination of a long and difficult path. For many years, an international consortium of scientists, mostly in the United States and the United Kingdom, and funded by government sources or biomedical charities such as the Wellcome Trust, had made slow but steady progress, releasing bits of sequence as they went along. They were called the public consortium because they received substantial public funding and had pledged to make their data available to all.

Then, in the early 1990s, J. Craig Venter, who had made his name by producing the first complete sequence of a bacterium, *Haemophilus influenzae*, entered the fray. Venter was something of a maverick in the field. He played the part of the American entrepreneur and capitalist, sailing around the world in his yacht, often flying by private jet. On one of the few occasions I saw him, he jetted into a meeting at the Cold Spring Harbor Laboratory to celebrate the 150th anniversary of Darwin's *On the Origin of Species*, gave his talk, and left immediately because he clearly must have had more important things to do—unlike me, who stayed for the rest of the weeklong conference. Venter had already caused a huge fracas in the science community when he worked at the U.S. National Institutes of Health (NIH)—the large government biomedical research laboratories in Bethesda, Maryland—by attempting to patent pieces of human DNA sequences to allow their commercial exploitation for treatment and diagnosis. The decision by NIH to green-light this led James Watson to resign as the first director of the agency's National Center for Human Genome Research. Although the NIH had filed the patents in his name, Venter said later that he was always against them.

Venter felt that the public consortium was too slow and that the method he had used for sequencing the million bases of a bacterium could be scaled up to sequence the roughly 3 billion bases in the human genome at much lower cost. So he started a private company, Celera, to do just that. Of course, Venter wasn't above using the large portions of the human genome that had already been sequenced by the public consortium before he entered the race. Many in the human genome community were outraged by Venter's audacity and were determined to ensure that the human genome, and, indeed, all other natural genomes, were not patented for the benefit of a private company but freely available to humanity.

One detractor was John Sulston, one of the leaders of the public consortium. Sulston presented a marked contrast to Venter. Despite his considerable fame and influence, the British scientist continued to dress in the sandals and other shabby attire reminiscent of a 1960s hippie. He lived in the same modest house and commuted to his lab on his ancient bicycle. A particularly passionate advocate of the genome being free for use by all, Sulston was sharply critical of Venter's motives and contributions. In the run-up to the completion of the draft sequence, relations between members of the public consortium and Venter became so acrimonious that President Clinton had to intervene personally to get them to politely share the stage at the announcement.

Despite all the hoopla, the draft sequence that Clinton and Blair announced was just the beginning. Large sections of the genome were still missing, especially regions consisting of repeating letters and thus difficult to sequence, and scientists had to figure out how some stretches of DNA actually fit together. The sequence was declared finished three years later, although, in reality, even today a few gaps remain, including on the Y chromosome, the male sex chromosome. (Women have two X chromosomes; men, one X and one Y.)

The human genome sequence is often called "the book of life," but this is somewhat misleading. In reality, even a perfectly complete sequence would be more like one long unpunctuated stream of text than a book. It would have no markings to denote individual chapters, paragraphs, or even sentences, nor cross-references to provide context. It would certainly be nothing at all like a well-edited encyclopedia in which you could look up your favorite gene and learn all about it and its relationship to everything else. And frankly, a lot of it was indecipherable. Only about 2 percent of our DNA actually codes for the proteins that carry out much of life's functions. The rest consists of what biologists once dismissed as "junk DNA"; they now increasingly think it *is* important, but don't fully understand how or why.

Initially, scientists didn't even know where a lot of the protein-coding genes were, because the signals that indicate where a gene starts and ends on the DNA are not always obvious. They are made even harder to discern by the presence of what are called pseudogenes: regions that once might have coded for proteins but are no longer expressed or functional. Many pseudogenes originated from viruses that inserted their own genes into our DNA. Finally, even knowing the sequence of a gene does not automatically reveal its function. Nevertheless, sequencing the genome was an immensely useful start. It allowed us to ask questions and conduct experiments that would have been unthinkable before. It was a watershed in biology.

You might also think that the book of life would be able to tell us accurately how each of our individual stories develops and ultimately ends. After all, DNA is the carrier of all genetic information, the master controller that oversees biological processes. Shouldn't knowing its entire sequence enable us to predict how an organism or cell will develop? Certainly mutations in individual genes have been associated with many diseases; examples include cystic fibrosis, breast cancer, Tay-Sachs disease, and sickle-cell anemia. But on the whole, biology is just not that deterministic.

Identical twins belie the view of DNA as destiny. They share the same genes and are often strikingly similar even when separated at birth. That's not surprising. What is surprising is that identical twins raised in the same environment can sometimes be *very* different, even when it comes to conditions with a strong genetic basis, such as schizophrenia.

Every one of us is a living testament to the fact that DNA by itself does not determine fate. All of our cells are descended from a single cell, the fertilized egg, and as that cell divides, it produces new cells, each one containing the same genes. Yet these genes give rise to a multitude of different cells. A skin cell is very different from a neuron, or a muscle cell, or a white blood cell. As we know, different genes are turned on and off in response to changes in the environment. It makes sense, then, that as different cells find themselves in slightly different circumstances, they change which genes they express and go down different paths to form the various tissues in the body. Importantly, you cannot reverse this process— even if you try to culture these different cells in exactly the same medium, they maintain their identity, as though the cells still remember which tissue they came from.

This suggests that some more permanent change has occurred in the genetic program of the cells as a result of their environment. The study of this change is known as epigenetics, from the Greek prefix *epi-*, for "above," to imply there was a second layer of control on top of our genes. The term was coined by the British polymath and professor of animal genetics Conrad Waddington in 1942. Waddington described the process in terms of a landscape. The original fertilized egg, he said, was like a ball on top of a mountain. Its progeny rolled down different paths into the various ravines and valleys at the foot of the mountain, each valley representing a different type of cell. Once there, it would be impossible to roll back up to the top or to roll up the ridge and down into a neighboring valley. In other words, once a cell had settled down into its final type, it couldn't change into a different type; a skin cell could not become a lymphocyte, a type of white blood cell. Nor could a skin cell reverse its fate and become a fertilized egg to give rise to an entirely new body.

Initially, Waddington was vilified by many as a Lamarckian, or someone who, like the evolutionary biologist Lamarck, believed that acquired characteristics could be inherited, an idea discredited by Darwin and Wallace's theory of evolution by natural selection. Waddington's theory seemed to imply that our environment affected our genes in some irreversible way. Even for those who accepted his ideas, they raised questions. At what point did the cell have its genome so altered that it could no longer direct the development of an entire organism? And how far down Waddington's mountain could a ball roll and still somehow go back to the top?

During Waddington's time, we did not even know that DNA was the genetic material, let alone its structure or how it stored genetic information. But it *was* known already that the fertilized egg, or zygote, was a very special cell: it had the right genetic material, and its cytoplasm, the internal material of the cell, seemed to have everything needed for kick-starting the process of developing into a new organism. The fertilized egg is said to be *totipotent*, meaning that it can develop into all the cell types needed to make a new animal, including its body and placenta. After a few divisions, the embryo reaches a stage called the blastocyst, which has a couple of hundred cells surrounding a fluid-filled cavity. The outer cells go on to form the placental sac, while the inner cells develop into everything else that forms the new animal. Those inner cells that develop into every cell in the body are called *pluripotent*.

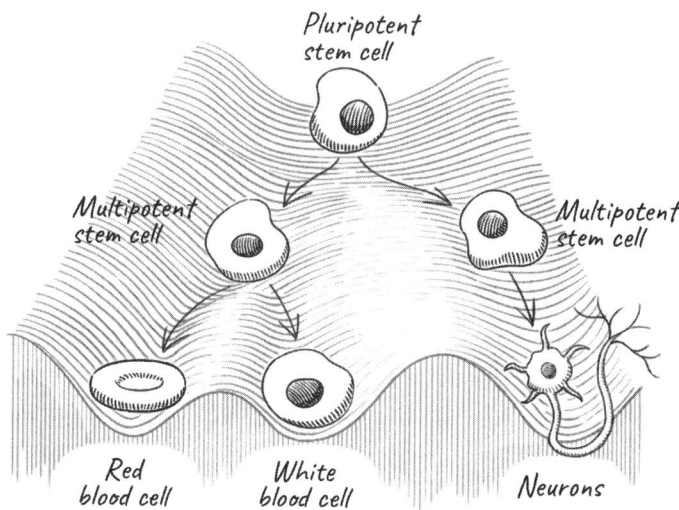

Waddington's metaphorical mountain shows the development of special cell types from a pluripotent stem cell.

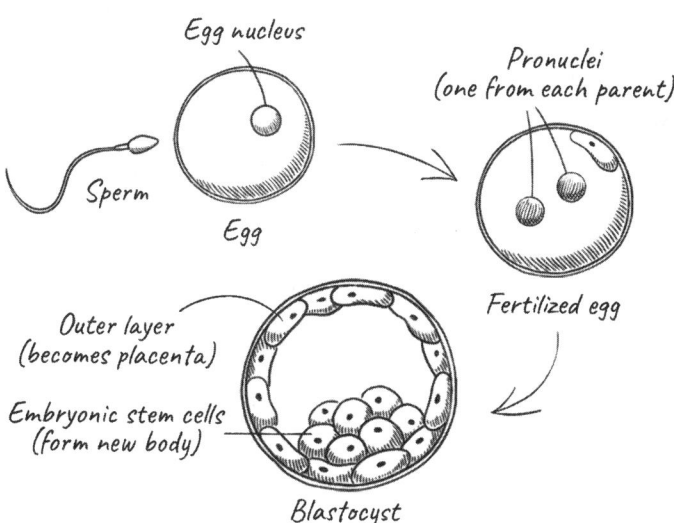

Development of a blastocyst from the fertilization of an egg.

Was the special property of the fertilized egg a result of its genome or its environment? If the latter, could you take a nucleus containing the genes from a highly specialized cell, put it into an egg that had its own nucleus removed, and make it totipotent so that it developed into a normal animal? This was precisely the question that Robert Briggs and Thomas King at the Institute for Cancer Research and Lankenau Hospital Research Institute in Philadelphia sought to answer. In 1952 they tried this with the northern leopard frog (*Rana pipiens*), as frog eggs are large and transparent, and thus easy to manipulate under a microscope. Briggs and King found that if they took nuclei from cells in the blastocyst stage of the embryo and introduced them into enucleated eggs, the eggs could develop normally into tadpoles. But if they took nuclei from cells at a later stage of development, the egg would develop partly and then stop and die. By a relatively early stage of development, then, an embryo's cells are already committed to their program. They are too far down Waddington's metaphorical hill and can't go all the way back to the top.

At this time, scientists simply did not know whether specialized cells had lost parts of their genome that were essential for growing an entire animal from scratch, or whether there was something else about them that prevented their development beyond a certain stage. Then along came a young scientist who would carry out one of the most famous experiments in modern biology.

WHEN I FIRST MET JOHN GURDON, I was immediately struck by his shock of golden hair that gave him a leonine appearance. By then, he was a world-renowned scientist in his seventies who worked in the institute named after him in central Cambridge, England, about three miles from my lab. Despite his stature in the world of science, he was unassuming and courteous to everyone, from a beginning graduate student to his senior colleagues. Long after many scientists would have retired, Gurdon remained passionate about science and carried out his own experiments. But his career had a rocky start.

Gurdon hailed from an aristocratic family whose Norman ancestor came with William the Conqueror in the 1066 invasion of England. Like many boys from privileged families, he went to Eton, the prestigious boarding school, at the age of thirteen. His time there did not begin well, for his biology teacher wrote a damning report at the end of his first science course. With the random capitalization that was already a couple of centuries out of date except in certain quarters of the British establishment, it said, "I believe he has ideas about becoming a Scientist; on his present showing, this is quite ridiculous, if he can't learn simple Biological facts he would have no chance of doing the work of a Specialist, and it would be sheer waste of time, both on his part, and those who have to teach him." Gurdon was not allowed to take any more science courses. He studied languages instead.

Nevertheless, Gurdon had a strong interest in biology and nature from childhood and was not so easily dissuaded. Fortunately for science, his parents were supportive and able to help him. Although they had already forked out several years' worth of expensive tuition fees to Eton, they paid for him to study biology with a private tutor for an additional year after he had graduated. In an unusual arrangement, he was then admitted to the University of Oxford on the condition that he first pass exams in basic physics, chemistry, and biology in a preliminary year. Gurdon survived the ordeal, began his undergraduate studies in zoology, and went on to begin research for a PhD with Michael Fischberg, who was also at Oxford. This was just four years after Briggs and King's experiment with frogs.

Fischberg suggested that Gurdon try to repeat their experiment but using a different kind of amphibian: the African clawed frog

(*Xenopus laevis*). Referred to originally as a toad, it was first brought to the attention of biologists by Lancelot Hogben, a peripatetic British scientist who moved from England to Canada and then, in 1927, became a professor at the University of Cape Town in South Africa. While there, Hogben began studying the frog because of its chameleonlike properties. The clawed frog became a favorite model organism in embryology; not only were its eggs large like those of the frogs that Briggs and King had studied, but also it had a short life cycle and could be triggered by external hormones to lay eggs any time of the year.

After overcoming some technical difficulties, Gurdon finally pulled off an experiment using *Xenopus laevis* that would revolutionize the world of biology. He was able to take the nucleus from one of the cells lining the intestine of a tadpole and insert it into an egg whose own nucleus had been inactivated by subjecting it to a large dose of UV radiation. The resulting egg developed into a complete tadpole, suggesting that the intestinal cell nucleus had all of the information needed for development that an egg nucleus had. To rule out the possibility that the egg's own nucleus had not been completely inactivated, Gurdon was careful to use two distinguishable strains of *Xenopus* for the cell that donated the nucleus and the egg that received it. There was no doubt that the donor nucleus had given rise to the tadpole. In fact, since the genes of the new tadpole were identical to those of the donor that contributed the nucleus, it was a clone of the parent. This was the first time that someone had taken the nucleus from the cell of a fully developed animal to clone an entirely new animal.

Gurdon's work had a tremendous impact almost immediately. He had demonstrated that the nucleus of a somatic cell of a fully developed animal was capable of directing the development of an entirely new animal—which would be a clone of the animal that donated the nucleus. It meant that a somatic cell could be made to go backward in development; in fact, all the way back to the top of Waddington's mountain. It could reverse the aging clock and start all over again to grow into a new animal. It also meant that cells that had developed into specialized tissues such as intestines retained all their genes. They were specialized not because they had preferentially lost genes but because they had somehow modified which genes would be turned on or off in each case.

Eventually other researchers reproduced Gurdon's experiments with different species, but the procedure was not performed on mammals until 1996. Scientists at the Roslin Institute, outside Edinburgh, cloned a sheep named Dolly from a cell taken from the mammary gland of an adult animal. The news generated huge headlines around the world. There was widespread discussion of the ethics of cloning, with concerns ranging from animal welfare to a brave new world in which rich people who wanted to live on would clone themselves or a loved one they had lost. (Apparently the absurdity inherent in this was also lost.) Today cloning has been successful in a wide range of animals, although for obvious ethical reasons, it is internationally forbidden to attempt it in humans.

In spite of all the excitement, Gurdon's early experiments were quite inefficient: only a small fraction of the nuclear transplantations actually worked. Others failed right away or developed into defective embryos that stopped growing and died. And in the sixty years since Gurdon's original experiments and the more than twenty-five years since Dolly, scientists have toiled painstakingly to improve the efficiency of cloning; nevertheless, it remains an inefficient technique. Nature's way of creating offspring works far better.

ONE OF THE BIG PROBLEMS with being human as opposed to, say, a starfish, is that we cannot generally regenerate our tissues. We cannot grow a new arm if one gets cut off. Soon after the first nuclear transplantation experiments, scientists began wondering whether the following might be the solution: Could you make these early embryonic cells grow on command into any type of tissue you wanted, such as heart muscle, neurons, or pancreatic cells? If that ever became a practical option, it would have enormous potential for medicine. Moreover, the deterioration of our tissues is one of the major problems we face as we age, and you could think of regenerating and rejuvenating them.

We might not be able to regrow a limb, but we already have the ability to regenerate certain kinds of tissue. Every time you cut or scrape yourself, your body creates new skin. Donate blood, and your body simply makes more. How does the body do this? While many of our cells are what we call terminally differentiated—they have reached a final state and will simply carry out their assigned tasks until they die—other, highly specialized cells are responsible for producing new cells to regenerate aging tissues. We call them stem cells.

Stem cells can be at many stages themselves. Many of them are already quite a way down Waddington's mountain, capable of developing into only a few different cell types. For example, hematopoietic stem cells in our bone marrow can generate all the major cells in our blood, including red blood cells and the cells of our immune system. But they can't become liver cells or heart muscle cells. However, the inner cells of the early embryo are pluripotent stem cells that can develop into every cell type in the body.

Scientists have been able to take these embryonic stem cells, or ES cells, maintain them in culture, and then alter conditions to nudge them into developing into one tissue type or another. Being able to grow ES cells in culture solved the problem of having to extract them from fresh embryos each time and fueled an explosive growth in stem cell research. However, the ultimate source of ES cells was still embryos, which would often be obtained from aborted fetuses, raising ethical questions and regulatory scrutiny. For some time, federal grants in the US could not be used to pay for research involving human ES cells, and labs had to clearly separate areas that were federally funded from those that were not.

It seemed almost miraculous that you could take any adult cell and coax it into developing into any tissue you wanted, let alone into an entirely new animal. What is it about stem cells, especially pluripotent stem cells, that makes them different from most cells in our body?

Molecular biologists had begun to identify transcription factors: proteins that regulate gene expression—that is, turning genes on or off, and by how much. The name comes from their control over whether a particular gene on DNA is "transcribed" into mRNA, which is then read to make the appropriate protein. Stem cells contained a large number of active transcription factors, some of which were needed to keep them growing in the laboratory. It was hypothesized that perhaps a newly fertilized egg possessed similar transcription factors that allowed it to develop into a new animal. Some of these same factors were also active in cancer cells, which can proliferate indefinitely.

Such was the state of affairs in the late 1990s, when a Japanese scientist, Shinya Yamanaka, turned his attention to the matter. Yamanaka was born in 1962, the same year as John Gurdon's successful cloning of a frog. He began his career as a surgeon, influenced partly by his father, an engineer who ran a small factory in the city of Higashi-Osaka. Yamanaka's enthusiasm for surgery soon waned, however: not only did he begin to lose confidence in his skills but also he came to see surgery as limited in terms of being able to treat many patients with intractable conditions such as rheumatoid arthritis and spinal cord injuries. Instead, Yamanaka thought, he ought to spend his life working as a basic scientist to find ways to cure them. He earned a PhD in Osaka and went on to postdoctoral research at the Gladstone Institute of Cardiovascular Diseases in San Francisco.

By the time Yamanaka returned to Japan to establish his own lab in the late 1990s, scientists knew that ES cells expressed quite a few transcription factors. If you turned on some or all of these factors in a normal cell, would you be able to trick it into behaving like a stem cell? Yamanaka and his student Kazutoshi Takahashi hoped so. They identified twenty-four factors that might be responsible for the pluripotent property of ES cells, and systematically introduced them into fibroblast cells found in skin and connective tissue—the same cells that Hayflick had attempted to culture. By experimenting with transcription factors in various combinations, they found that just four were enough to convert an adult fibroblast cell into a pluripotent cell.

As a result of Yamanaka's work, we no longer need to harvest cells from embryos to generate pluripotent cells; we can make them from other adult cells. The pluripotent cells made using Yamanaka factors are called induced pluripotent cells or iPS cells. The increased ease of generating iPS cells has led to an even greater explosion in the field of stem cells. Scientists are constantly improving both the efficiency and safety of the process, as well as becoming increasingly sophisticated in determining the paths that the stem cells can take.

REMARKABLE AS THESE ADVANCES ARE, they don't tell us exactly what is happening to our genome that makes cells behave so differently even though they all have the same DNA. Why do different cells have such different genetic programs? And why do cells remain true to type, so that one cell type doesn't suddenly change into a different one? Even stem cells that are responsible for generating blood cells don't start producing neurons or skin cells.

Each cell carries genes that are always expressed because every cell needs them. They're referred to as housekeeping genes. But for other genes, which ones are turned on and which are kept switched off depends very much on what that particular cell needs. How does the cell control this process? You just read about transcription factors, proteins that control which genes are actively expressed or repressed. One of the first and simplest examples of such a factor was discovered in exploring how the bacterium *E. coli* digests the simple sugar lactose. Ordinarily, *E. coli* doesn't encounter lactose, so it does not constantly make the enzymes necessary to digest it. Instead, it operates on an as-needed basis: when the bacterium senses lactose, it turns on the genes tasked with turning out the appropriate enzymes. As soon as there is no more lactose around, it shuts down those genes. It is a simple and elegant way to switch genes on or off in response to a change in the environment. A good deal of gene regulation works exactly like that, by controlling transcription in response to a stimulus. It is seldom as simple as the lactose case, and usually involves a complicated network where genes that are activated in turn activate or switch off other genes, which affect even more genes.

With *E. coli*, you can reverse the response to lactose simply by removing lactose from the culture. But if you took a skin cell and put it into, say, a liver, it wouldn't suddenly start behaving like a liver cell. The transcription factors of a skin cell and a liver cell are different; in addition, the cell has a way of ensuring that some changes in the genetic program persist for a long time, which involves rewiring the code on DNA itself.

So far, we have thought of DNA as a simple four-letter script containing all the information to make the proteins that carry out various essential functions. But even before the structure of DNA was known, scientists understood that a small fraction of its four bases, A, T, C, and G (or U, the equivalent of T in RNA), had extra chemical groups attached to the base. In the early days, nobody knew what these modifications were for.

Today we know that many of them act as extra tags that serve as signals for whether a gene should be kept switched on or off over the longer term. The most common of these is the addition of methyl ($-CH_3$) group to cytosine, the C base in DNA. When Cs at the right place are methylated in this way, the genes just ahead of them are kept switched off.

As cells develop, they will methylate their DNA in the region of genes they want to shut down, and leave unmethylated those regions that contain genes they need to actively use. So cells that differentiate into skin cells will have a different methylation pattern from, say, neurons.

You might expect that when cells divide and their DNA copied, the patterns of methylation would be lost because you're making the new DNA with fresh building blocks, but the cell has an ingenious way of restoring the methylation pattern of the parent cell. What this means is that the exact pattern of methylation can be passed on to the daughter cell when a cell divides, so genes that are shut off in a particular cell lineage remain shut off. The flip side of this also occurs: there are demethylases that remove methyl groups, which then allow those genes to be turned back on. Apart from using transcription factors, modifying the DNA itself in this way offers a completely additional level of control over which genes are turned on and off. It is also a method of ensuring that these changes can be passed on to the next generation of cells. These modifications of DNA alter the way our genes are used. They are called epigenetic marks or changes because they are the molecular explanation for the phenomenon of epigenetics that Conrad Waddington had first described.

These epigenetic marks not only persist and even increase as we age—they can even be passed across generations. Toward the end of World War II, between September 1944 and May 1945, the Netherlands suffered from a devastating famine that would claim the lives of more than 20,000 people. A later study showed that despite the relatively brief duration of the famine, the children of women who were pregnant during the mass starvation suffered adverse physical and mental health consequences throughout their lives. They experienced higher rates of obesity, diabetes, and schizophrenia, and had a higher mortality than children who were not in utero during the famine. The effects were even different depending on whether the famine occurred in the early or late stages of pregnancy. Comparing the DNA of subjects who had experienced starvation in utero with those of their older and younger siblings was revealing: the famine had imposed on the fetus a methylation pattern that had consequences over the course of its life and accelerated both aging-related diseases and mortality. It is a striking example of how an external stress can cause epigenetic changes to DNA that last a lifetime.

IF THAT ISN'T COMPLICATED ENOUGH for you, just wait: DNA isn't present in cells as a naked molecule. Rather, it is heavily coated with proteins called histones, and this mixture of proteins and DNA is called chromatin. These histones help us understand how all of our DNA can fit into a cell's tiny nucleus. If you could stretch out the DNA in a cell, it would measure approximately two meters (six and a half feet). The nucleus, in contrast, is only microns in diameter—or about a million times smaller. Histones are positively charged and neutralize the negative charges on the phosphate groups of the DNA. By doing so, they allow DNA to condense into a highly compacted form.

The first level of DNA compaction is the nucleosome, in which DNA is wound around a ball-like core consisting of eight histone proteins. The nucleosomes further organize themselves into filaments that are then woven back and forth until it all fits comfortably in the nucleus. When cells divide, the duplicated chromosomes have to move into each daughter cell, and just as you would cram the belongings from your entire household into a truck before you move, chromosomes are most compact just before cell division. That is when they have the familiar X shape that we see in most popular images of chromosomes. But for most of the life of the cell, chromatin is much more extended.

The problem with compacting chromatin is that the cell needs to be able to access information on the DNA when needed. It's like owning a large collection of books but not having sufficient space in your home to have all of them within easy reach. You might box most of them and store them in the attic but keep the books you're currently reading or planning to read soon easily accessible on a bookshelf or piled on your nightstand. The cell too has to make sure that appropriate regions of chromatin are accessible, even if it wants to shut down much of it. It does so by tagging histones by adding certain chemical groups to them. Just as with methyl groups on DNA, there are enzymes that add these histone tags and others that take them off. Tags on histones can act as a signal for the cell to recruit other proteins to that region and either inactivate chromatin or open it up, so they too act as epigenetic marks. With histones, one common tag is called an acetyl group, and the enzymes that add them to histones are called histone acetylases.

In general, DNA methylation and histone acetylation exert opposite effects. DNA methylation usually silences the gene that follows the methylated region, while histone acetylation signals that the gene is to be actively transcribed. Both can be reversed by the action of demethylases or deacetylases.

What both modifications do is to overlay on top of the DNA sequence itself a second and longer-lasting way of modifying the program of a particular cell. They allow cells to maintain a stable identity as neurons, skin cells, or heart muscle cells. As a cell develops from the fertilized egg, different epigenetic marks must be laid down as it develops into different cell types.

WE ALL KNOW THAT PEOPLE age at different rates. Some people look old at fifty, while others are remarkably youthful into their eighties. Some of this comes down to genetics, but aging can also be accelerated by stress and hardship. From the moment we are conceived, our cells don't just acquire mutations in the DNA affecting the underlying code itself. They also acquire epigenetic marks. As we saw with the Dutch famine survivors, some of those marks are the result of environmental stress.

Steve Horvath, while working at the University of California, Los Angeles, was not interested in epigenetics, believing it to be too messy, indirect, and unlikely to show much useful connection to aging. But one day, a colleague was collecting saliva from identical twins who differed in sexual orientation, and he wanted Horvath to help him see if there were any epigenetic differences between them. Horvath is a twin; his brother is gay, while he is heterosexual. In the spirit of scientific inquiry, they contributed some of their own spit to the study. When they looked at the methylation of cytosines, they found absolutely no relationship between the pattern and sexual orientation.

But Horvath now had a lot of data from twins of various ages. He decided to mine it further to see what else he could learn. He discovered a very strong correlation between the DNA methylation pattern and age. He then looked at cells in other tissues and correlated the methylation pattern with actual markers of aging—for example, the sort of things your doctor would analyze from your blood, such as liver and kidney function. He was able to identify

513 sites of methylation that could predict not only mortality but also cancers, health span, and the risk of developing Alzheimer's disease.

These patterns help scientists approach a fundamental problem. People age biologically at different rates, so how do you measure aging? Methylation patterns are like a biological clock; in fact, they are more accurate than chronological age alone at predicting age-related diseases and mortality. Many other research groups developed their own methylation clocks with slightly different markers, all correlating well with biological age. Still, as Horvath and his colleagues themselves point out, these clocks are useful for research but are not yet a substitute for tests that measure loss of physiological function or provide early diagnosis of diseases.

We don't think of young children as aging; in fact, throughout much of childhood and adolescence, they become stronger and their odds of dying decline. But it turns out that while the methylation patterns reverse very early in the embryo, suggesting a resetting of the clock or a rejuvenation, from that point on, methylation follows an inexorable pattern. So we age from even before we are born! Similarly, the long-lived naked mole rat is thought not to age because its risk of dying doesn't increase with time. In fact, its methylation pattern shows that it does age, just more slowly than other rodents.

For an extreme example of the effect of epigenetics on longevity, look no further than a beehive. Bees, like ants, have a queen that can live many times longer than other bees that share exactly the same genes: queen honeybees live two to three years, while worker bees die after only about six weeks. This is partly because once the queen is selected, she is treated very differently. She is kept deep in the hive, pampered and protected against predators, whereas worker bees and ants must go out and risk their lives foraging for food. She is fed an exclusive diet of royal jelly, which has a different composition and a much higher nutritional value than the ordinary nectar and honey that worker bees live on. But the impact of these factors goes deeper. Something about her diet and stress-free environment results in her having different epigenetic marks from worker bees, and she ages at a far slower pace.

The question of why epigenetic marks should cause aging is complicated. The patterns are associated with an increase in inflammatory pathways and a decrease in pathways for making RNA and proteins as well as DNA repair, so it is easy to see how they might result in aging.

The epigenetic changes also seem to occur on a timetable. This doesn't mean that aging itself is programmed. It could simply be that the epigenetic changes take place when they are needed at some stage, but they are not switched off when their work is done because evolution doesn't care what happens to you after you have passed on your genes. By shutting down many genes in a stable way, epigenetics may also prevent cells from becoming cancerous early in life. Like telomere loss, and the response to DNA damage, this may be yet another example of the trade-off between preventing cancer and preventing aging.

It is also possible that many epigenetic changes are not programmed but caused by random changes in the environment. Remember the case of identical twins? Those epigenetic changes in their DNA diverge right from birth, so while they still have largely the same DNA sequence, they acquire very different epigenetic marks.

CAN THE AGING CLOCK EVER run backward? Yes, and it has happened to every single one of us: at conception, when the aging clock is reset to zero. When a forty-year-old woman gives birth, that newborn is not twenty years older than a baby born to a twenty-year-old woman. Even though the germ-line cells are older in the forty-year-old woman, both children start at the same age. The aging that takes place in the parents is reset in the child.

We have evolved at least three ways to reset the aging clock. The first is that germ-line cells have superior DNA repair and accumulate fewer mutations than somatic cells do.

Second: the egg and the sperm each undergo a rigorous selection process prior to fertilization. A woman produces all the eggs she will ever have while she is still a fetus. These number perhaps a few million to start with but are down to about a million by the time she is born. By puberty, this number drops to about a quarter million, and by the time a woman is thirty, only about 25,000 eggs remain. However, a mere 500 of those eggs get used up by ovulation during the menstrual cycle over a woman's lifetime. With sperm, this ratio is even more dramatic: males produce millions of sperm cells from puberty on. So there is a huge surplus of both eggs and sperm. Why? Prior to ovulation—that monthly event in which the ovary releases one mature egg, or ovum, into the fallopian tube for the purpose of potentially being fertilized—the eggs in the ovary are somehow inspected and destroyed if damage is detected. Only those that pass the test make it to ovulation. As damage is likely to increase with age, this might explain why the egg count drops precipitously and the chance of becoming pregnant decreases. Perhaps the monitoring process also becomes less effective, since genetic defects in the baby also increase with the age of the mother.

Similarly, sperm cells may undergo selection as well, and a sperm must swim and outcompete all the millions of others to be the first one to fertilize the egg. Even after fertilization, many embryos are rejected early in development if they are sensed as being defective. And even within an embryo that is developing normally overall, there is competition to eliminate abnormal cells. The process isn't perfect, but nature has done its best to ensure that our offspring are free of our own cellular damage and aging.

The third method for resetting the aging clock is to actually reprogram the genome. Immediately after impregnation, the fertilized ovum, or zygote, temporarily bears *two* nuclei (pronuclei): one from the mother and the other contributed by the father. The enzymes and chemicals in the zygote proceed to erase nearly all the epigenetic marks in the DNA of both pronuclei, and then add new ones to start the fertilized egg on the path to making a baby. Notice that I said "*nearly* all." An egg with both pronuclei coming from just a male or female parent alone would not develop normally. This is because the pronuclei donated by the mother and father have a different but complementary pattern of epigenetic marks, also called imprinting, which together provide the proper program for development.

Considering all the intricacies of normal development we just described, it is amazing that cloning frogs or Dolly the sheep ever worked at all. For one thing, the genome of cloned animals came from adult somatic cells, with an entire lifetime of accumulated damage. Animals conceived normally, on the other hand, start off from much more protected germ-line cells and go through a rigorous selection process both before and after fertilization. In addition, changing the program of a somatic cell is very different from an egg's normal task. Given these difficulties, how could these cloned animals possibly be normal? Would they not show signs of premature aging or other abnormalities compared with naturally conceived animals? In truth, it didn't work so well. Most of the transplants never made it to fully formed animals. Still some, like Dolly, did.

And the truth is, Dolly was quite a sick sheep. She had abnormally short telomeres and, at the age of one, was judged as older than her chronological age by several criteria. Sheep normally live ten to twelve years, but at six, poor Dolly developed tumors in her lungs and had to be put down. It turns out, however, that Dolly was not the only sheep cloned. There were also the lesser-known Daisy, Diana, Debbie, and Denise, who, surprisingly, all lived healthy lives with a normal life span. This suggests that, at least in principle, it may be possible to reverse the effects of aging and reset the clock even if you start from an adult somatic cell, just by reprogramming the cell. Erasing the epigenetic marks and initiating a new program of gene expression can enable a newly cloned animal to begin from scratch.

Cloning, though, is not the main aim of reprogramming cells, even for farm animals or crops. The real payoff would be in using stem cells for regenerative medicine: repairing or replacing tissue that has died or sustained damage. If we can overcome the technical problems, the possibilities are enormous and wide-ranging. Perhaps we could introduce new pancreatic cells that produce insulin in patients with diabetes, replace damaged heart muscles after a heart attack, or even regrow neurons in people who have suffered a stroke or a neurodegenerative disease like Alzheimer's. The potential for such breakthroughs is why billions of dollars are being invested in stem cell research today.

Even though they're not going all the way back to zero and creating a new cloned animal, these stem cells are effectively trying to reverse the aging clock by regenerating or even replacing individual parts of an animal that have aged. Both embryonic stem cells and induced pluripotent stem cells (iPS cells) are capable of differentiating into numerous cell types, but the two are not exactly the same. ES cells are natural early embryonic stem cells that scientists have figured out how to keep cultured and then program to follow different paths to make different tissues, whereas iPS cells are reprogrammed not by the action of factors in the egg but by using the four Yamanaka factors in a somatic cell. This means their behavior is not exactly the same. Still, because of the convenience of generating iPS cells (without the added burden of having to contend with the legal and ethical issues surrounding ES cells), many scientists are working hard to improve Yamanaka's original method for reprogramming cells.

We will soon see how scientists are trying to reverse aging using this approach. There is also much interest in reprogramming the cell by using specific compounds that inhibit DNA methylation or histone deacetylases. This route to rejuvenating tissues, and even the whole animal, is a major focus of current research. As with telomerase, it may well be the case that our epigenetics have evolved to strike a fine balance between reducing the risk of cancer early in life and accelerating aging. Thus, any approaches to slow down aging or attempt to reverse it by rejuvenation may have to contend with how to do it safely. Indeed, many tissues that have been generated using the four Yamanaka factors have been associated with an unusually high proportion of tumors.

In the last three chapters, we have seen how the genetic program that controls life can be disrupted by damage to our genome, accumulated with age. We have seen how the program itself is modified on the fly to suit the organism's needs at any given stage. The product of the program is the ensemble of proteins in our cells. These proteins carry out a huge number of complex and interconnected tasks and are like players in a large symphony orchestra.

Now we will see what happens when that orchestra becomes discordant and breaks down.

6.

RECYCLING

THE GARBAGE

These days, whenever I forget an appointment or misplace my gloves, umbrella, or hat, I panic for a moment. I have just turned seventy as I write this, and these occurrences immediately strike me as signs of an inevitable and worsening decline. I cheer up when I remember that in my early twenties, I once invited a friend to dinner, forgot about it, and wasn't even home when he called; or that a couple of years later, I was so preoccupied with finishing my work that I forgot to attend my own going-away party that a neighbor was going to throw for me. And that I've been notorious for losing things all my life.

Still, there is a good reason for my foreboding. We all face the prospect of suffering from neurodegenerative diseases that cause us not just to forget but also to completely lose our sense of who we are.

Today more than 50 million people suffer from dementia, and as the proportion of older people in the population is increasing in almost every country in the world, that number is expected to grow to 78 million by 2030 and 139 million by 2050. In England and Wales, it recently overtook heart disease as the leading cause of death, partly because treatment of heart disease has vastly improved, while there is still no effective treatment for dementia. In the United States, it still lags behind the more established killers such as heart disease, cancer, and accidents, but its proportion is gradually rising. It is estimated that about one-third of people born in 2015 will go on to suffer from some form of dementia.

Over half of those with dementia have Alzheimer's disease, named after the German psychiatrist Alois Alzheimer, who, around 1900, characterized the onset of the then-unnamed disease. His patients, he wrote, would oscillate from periods of calm and lucidity to being unable to identify common objects, feeling increasingly disoriented, forgetful, agitated, and even unhinged. That is just the beginning. As the disease progresses, many Alzheimer's sufferers are unable to recognize their family and friends. They can no longer carry out basic activities such as speaking, eating, and drinking. They become increasingly terrified at their loss of control, their loss of self-identity, and their increasing inability to make sense of the world around them. Their loved ones may have it even worse, though, having to watch this person—a spouse, a grandparent, a cherished friend—gradually vanish.

In the century-plus since Dr. Alzheimer's description, we have made tremendous progress in understanding the biology behind Alzheimer's disease. The same is true of other neurodegenerative maladies, such as Parkinson's and Pick's diseases. They all have two things in common: the likelihood of the disease increases as we grow older; and they are caused by a malfunction of our own proteins.

Proteins, as we have seen, are long chains of amino acids that miraculously fold up as they are made. Well, not miraculously. The reason that they fold up is that some amino acids, like oils, are hydrophobic, meaning that they do not like to be exposed to water. Hydrophilic amino acids, on the other hand, are happy to interact with water molecules. As a protein chain emerges, it folds into its characteristic shape by tucking away most of the hydrophobic amino acids on the inside of the protein and exposing the hydrophilic ones on the outside where they are in contact with the surrounding water. Most protein chains have a particular shape or fold that is stable and functional. Sometimes a protein chain folds up along with others to form a complex of several chains. But the principle is the same. In an amazing display of coordination, each of our cells makes not one but thousands of proteins in the amounts it needs and at the time it needs them, and they all must work together as a well-orchestrated ensemble. But the process can, of course, go wrong.

Think of the many ways a household item can become useless. Even a brand-new product can be poorly made and arrive saddled with manufacturing defects. You could damage it accidentally while using it. Or it could slowly wear out or rust and become dangerous to use or stop working entirely. Then there are products, once essential, that we no longer need. Perhaps our children have grown up, and we no longer require baby bottles or cribs. Or technology has changed, and we have no use for a cassette recorder or a film camera. Or our possessions simply go out of style, and we no longer want them. Food has an even shorter shelf life. In our daily lives, we deal with all this as a matter of course. We throw out leftover food that has perished, mend or throw out old clothes, and fix or get rid of broken gadgets. If we didn't do that, our homes would quickly fill up with junk and become unlivable.

It is the same with cells and their proteins. Proteins can have manufacturing defects too. The protein chain may be made incorrectly or be incomplete. It might not have folded into its appropriate shape. During its lifetime, it could lose its shape by unfolding or be damaged by chemicals or other agents. Just as we may need items only during a particular phase in our lives, many proteins are needed only briefly at a particular stage during a cell's development or in response to some environmental stimulus. And just as we dispose of or recycle products that are faulty or have simply worn out or been damaged, the cell has evolved ways to detect and then destroy proteins that are defective to begin with or when they become aberrant later. It also has ways of getting rid of perfectly normal proteins that it no longer needs. In all these cases, the cell breaks down defective proteins into their amino acid building blocks, which it can then use to make new proteins or to produce energy.

However, there are crucial differences between the proteins in a cell and a home full of household items. Manufacturers don't usually much care what happens to their products after they are sold (except during the warranty period, of course). Moreover, the manufacturer of your washing machine does not have to make it compatible with other appliances and therefore isn't concerned about which brand of refrigerator or microwave oven you own, or whether you own one at all. Cells, on the other hand, both manufacture proteins and use them, and have to ensure that the many thousands of proteins all work together without problems.

As we age, the quality control and recycling machinery of the cell deteriorates, leading not only to neurodegenerative but also many other diseases of old age, including inflammation, osteoarthritis, and cancer. Accordingly, the cell has come up with multiple ways of ensuring the quality and integrity of its collection of proteins.

Proteins can be defective in many ways. The birth of a protein chain takes place on the ribosome, the large molecular machine that I have studied for the last forty-five years. As the ribosome chugs along, it reads the genetic instructions on mRNA to stitch together amino acids in a precise order to make a protein chain. The process has evolved to a high level of perfection over billions of years, but it still occasionally gives rise to defective products. Sometimes the mRNA contains mistakes; sometimes the ribosome misreads it. In these cases, the newly made protein has the wrong sequence of amino acids, so it malfunctions—a bit like a brand-new gadget with a manufacturing defect. These days, many of my colleagues and I are trying to understand how the cell recognizes these mistakes and homes in on them for removal.

Even if the new protein chain has the correct sequence of amino acids, as it emerges from a tunnel in the ribosome, it still faces the challenge of folding into its proper shape. Although the protein chain contains within it all the information needed to form that shape, the process doesn't usually work spontaneously. With larger proteins, it is difficult to keep the hydrophobic sections from different parts of the chain apart so that they do not stick to one another (or even worse, to other chains that are being made at the same time) while the protein is folding. There are many ways that the folding process can go awry, so cells ranging from bacteria to humans have evolved special proteins whose purpose is to assist other proteins to fold correctly. Ron Laskey, one of my fellow scientists in Cambridge, humorously named these proteins chaperones. (Among other things, Laskey is a folk singer who has written and recorded witty songs about life as a scientist. One of his songs is about how, as a young man, he was part of a double bill with Paul Simon in a small venue in England when neither of them was well known—and realized immediately that he had better stick to science.) Like Victorian chaperones during courtship, these proteins prevent improper interactions between different parts of the chain or between chains. Even so, proteins occasionally misfold.

Even after a protein has already folded into the right shape, you can make it unfold. The proteins in a chicken egg are all folded correctly to carry out their collective function of helping a fertilized egg grow into a chick. But if you take that egg and boil it, its proteins unfold. Similarly, if you add lemon juice to milk and stir, the acid unravels the proteins in the milk. In either case, when the protein chains unfold, the water-avoiding hydrophobic amino acids that were on the inside now become exposed to the surrounding liquid. This makes the proteins stick to one another and become tangled, and the egg or milk turns into a gelatinous solid.

Even without being boiled or treated with acidic lemon juice, proteins are not rocklike, static entities. The atoms in a protein jiggle around all the time, and the proteins themselves breathe and oscillate around their average shapes. Over time, they can unfold, either spontaneously or in response to environmental stress. Often the proteins will then fold back into their original shapes, but sometimes they will clump together instead. As we age, more clumps means more proteins that have lost their function. Even more seriously, the protein aggregates themselves can lead to diseases such as dementia.

We can thus have proteins that are incorrectly made to begin with, or proteins that misfold later. But that's not all. Many proteins have extra sugar molecules added to specific points on their surface after they are made. This process, called glycosylation, is essential for their work. But as we age, sugar molecules are added randomly to proteins, a process called glycation, to distinguish it from the normal and orderly process of glycosylation. Glycation causes a number of common health problems. For instance, eye diseases such as cataracts and macular degeneration result from proteins in the lens or retina of our eye being modified by sugar molecules, which changes their properties and prevents them from functioning normally. These proteins too need to be recognized and destroyed before they become a problem.

The first line of defense are the chaperones, which refold misshapen proteins into their correct shapes. But if unfolded proteins accumulate, more drastic action becomes necessary. Cells have an elaborate sensor to detect the buildup of unfolded proteins. The unfolded protein response, as this is known, is multipronged: First, more chaperones are synthesized to aid in folding these aberrant proteins. Second, they are tagged and targeted for destruction. Since there is clearly a problem with proteins folding properly, the cell also slows down protein production or shuts it down entirely. In extreme cases, where these measures are inadequate, the unfolded protein response can simply direct the cell to commit suicide.

How can a cell destroy proteins that it senses as defective or unwanted? When it senses that something is wrong, it tags the protein with a molecule called ubiquitin, which is itself a small protein. Ubiquitin was discovered in the mid-1970s and got its name from the fact that it was ubiquitous—scientists found it in almost every tissue they examined. It seemed to have something to do with regulating proteins in the cell, but exactly how wasn't clear.

Eventually researchers discovered a huge molecular machine called the proteasome, which acts as a giant garbage disposal. When a ubiquitin-tagged protein is fed into the proteasome, it gets chopped up into pieces that can be recycled. Of course, you can imagine that such a powerful degrading machine could be quite dangerous if it were free to act on proteins at will. So the entire process is highly regulated. It is used not just for defective proteins but also for perfectly functional proteins that are no longer required.

Any defect in the proteasome or the ubiquitin tagging system means that unwanted proteins hang around the cell and cause problems. Proteasome activity declines with age, and we have reason to believe it is a cause of aging. Deliberately introducing defects in the proteasome or the ubiquitin tagging machinery can be lethal, and even minor defects can lead to diseases associated with old age, such as Alzheimer's and Parkinson's.

The ubiquitin-proteasome system is beautifully tuned to get rid of unwanted or aberrant proteins. It works by chewing away the strand of a single protein at any given time. Like the garbage disposal in your kitchen sink, it can handle only one scrap at a time. But what if a cell wanted to get rid of a lot of very large junk, much as we would want to get rid of a used sofa, old furniture, or appliances? Not to worry. Nature has this covered with an apparatus that, oddly enough, was discovered decades before the proteasome.

Scientists have long known that cells from higher organisms have a nucleus that contains our chromosomes, but as they studied the cell in greater detail with ever more powerful microscopes, they discovered that they have many other specialized structures called organelles. How these structures worked together to facilitate cell function remained a mystery. One of those structures turned out to be hugely important for recycling the cell's garbage.

In 1955, Christian de Duve, who split his time between Rockefeller University in New York and the Catholic University of Leuven in Belgium, discovered an organelle called the lysosome. He and his Leuven colleagues found they were full of digestive enzymes that would break down any of the major constituents of living matter. Initially the lysosome was considered rather boring—about as exciting as a landfill site in a city. But things became more interesting when scientists showed that lysosomes often contained remnants of other parts of the cell. All kinds of unwanted structures were taken to lysosomes for disposal. De Duve coined the term *autophagy*, from the Greek for "self-eating," because the cell was digesting away parts of itself. But how did the cell's garbage make its way to the lysosomes?

In the cell, membranous structures called autophagosomes form and grow in size, gradually engulfing everything the cell targets for disposal. Think of autophagosomes as large garbage trucks. The garbage they collect can be anything from protein aggregates all the way to large organelles. An autophagosome eventually merges with a lysosome to deliver its contents to be digested and recycled. If the proteasome is akin to the garbage disposal in your kitchen sink, the lysosome is the huge garbage recycling center in your city.

While this process goes on perpetually, it is highly regulated. If you stress or starve the cell, autophagy goes up. It makes sense to break down proteins and other structures and recycle their components to survive a difficult time.

However, this still doesn't tell us how the cell decides when and what to deliver to lysosomes. Science would have to wait almost fifty years to make headway on this problem. In the late 1980s and early 1990s, Yoshinori Ohsumi, a young assistant professor at Tokyo University, hatched a clever idea.

Biology often advances by studying simple organisms that are easy to grow and mutate, and the discoveries made there can then easily be generalized to more complex ones such as humans. Ohsumi turned to that favorite of molecular biologists, baker's yeast, in which the equivalent of the lysosome is called a vacuole. By isolating strains in which the vacuole had accumulated cellular debris, he was able to find a dozen genes that were essential for activating autophagy.

As a result of these breakthroughs, we know now that autophagy happens continuously as part of the general maintenance of the cell. Its rate can go up or down, depending on the cell's needs. It can also be triggered when the cell needs to get rid of invading viruses or bacteria. This kind of autophagy requires special adaptor proteins that recognize these foreign objects and bring them to the auto-phagosome, which then delivers them to lysosomes to be destroyed. Autophagy is the only process by which the cell can destroy such enormous structures.

You might think that the only function of autophagy is to deal with problems, but it is also essential for a single fertilized egg's development into an adult animal. Imagine that you have a perfectly serviceable house, but you want to remodel it. Maybe you've had a new addition to your family, or you suddenly need more space so that you can work from home during a pandemic. Or you simply want a larger kitchen. When you remodel a structure, you have to break down parts of it before you can start building. You may have to take down walls, plumbing, and counters, or get rid of furniture that won't fit in the new space. Our cells go through this same process as they develop from that original fertilized egg into specialized cells such as neurons and muscles, which have very different internal organization and structures. Autophagy makes it happen.

In short, autophagy is used both to ensure cells develop normally and to jettison defective proteins or aging structures, as well as to destroy bacteria and viruses. It has so many essential functions that when it fails even partially, we develop serious problems, from cancer to neurodegenerative diseases.

So far, we have talked about how cells deal with proteins and larger structures that are defective or they don't need anymore. If there are just too many defective proteins piling up, it becomes hard for the recycling machinery to keep up. In that case, it would make sense to quickly shut down the synthesis of new proteins, a bit like turning off the main water supply when you have a flood in the bathroom. Also, it makes no sense for cells to produce new proteins and grow when they face starvation or stress.

One way the cell does this is to stop ribosomes from starting the process of reading mRNA to make proteins. It is a way of slowing down the production of new proteins while it handles crises, which is a bit like seeing a traffic jam on a freeway and preventing cars from entering the on-ramp and making the problem worse. While this process shuts down the production of most proteins, it also turns on the production of proteins that help the cell survive the stress and alleviate it. In the traffic jam analogy, this would be like sending a signal that stops new cars from entering the freeway and at the same time bringing in tow trucks to clear the accident that caused the jam.

This process of shutting down the synthesis of most proteins while allowing a few useful proteins to be made can be triggered by starvation, a viral infection, or too many unfolded proteins. Since it is a unified response to many kinds of stress, it is called the integrated stress response, or ISR.

You would think that these problems with protein quality and quantity would worsen with aging, making a strong ISR useful. That is exactly what some groups have found. If you delete the genes that turned on ISR in mice, the rodents were more prone to various pathologies caused by abnormal protein production. When mice suffering from a pathology due to unfolded proteins were treated with a compound that allowed ISR to persist, it alleviated their symptoms, whereas, conversely, suppressing ISR made them worse and hastened their demise. Compounds such as guanabenz or its derivative Sephin1 that strengthen the integrated stress responses prevent diseases caused by poor quality control of protein production. They also extend life span, although in at least one case, there was disagreement about how these compounds acted, and whether they even affected ISR directly.

If all this makes a strong case for restoring or strengthening ISR as we age, some research groups have found the exact opposite. According to their studies, deleting the genes that turn on ISR alleviated some of the symptoms of Alzheimer's disease in mice, including memory deficits. A molecule that shut down ISR enhances cognitive memory and reverses cognitive defects following traumatic injury to the brain. Even more surprisingly, the effects were seen even when the experimental drug being tested, an integrated stress response inhibitor—ISRIB, for short—was administered a month after the trauma.

Why would turning off a universal control mechanism be beneficial? Nahum Sonenberg, an expert on translation at McGill University in Montreal and a coauthor of the ISRIB study, believes there are pathological conditions in which the ISR itself is chronic and out of control. It may be suppressing protein synthesis when it shouldn't or to a much greater degree than it should. It's like driving a car in which the brake is activated all the time instead of only in response to a signal to slow down or an accident ahead. Instead of being a lifesaver, it becomes a nuisance. Even as we age, we still need to make new proteins. For example, forming new memories requires synthesizing new proteins that strengthen connections between brain cells. But when ISR is itself out of control, we are unable to make proteins in the amounts we need. In cases such as this, turning off ISR may be beneficial.

ISRIB has been touted in the press as a "miracle molecule" that could boost fading memory and treat brain injuries. The San Francisco company Calico Life Sciences, owned by Alphabet, the parent company of Google, started conducting clinical trials on ISRIB-like compounds that inactivated ISR. Peter Walter, one of the discoverers of the unfolded protein response and of ISRIB, recently gave up a prestigious professorship at the University of California, San Francisco, to join Altos Labs, a private company that operates research institutes to tackle aging, with campuses in California and Cambridge, England.

How this will play out is unclear. It is well to remember that ISR is a universal control mechanism precisely to deal with situations that are problematic for the cell, such as an accumulation of un-folded proteins, amino acid starvation, and viral infections. As we discussed above, initially, scientists found that prolonging ISR was beneficial for certain pathologies. So there may be situations when it would be helpful to enhance ISR and others in which it would be better to inhibit it. Figuring out exactly how much ISR is optimal at any given stage is unlikely to be straightforward, and we may have some way to go before it can be used with any confidence as a long-term treatment for combating diseases of aging.

We have covered a lot of ground in this chapter, but a common thread runs throughout. For cells to be able to function, their thousands of proteins have to work together. They must be produced at just the right time and in the right amount, and they must be the correct shapes. It is not unlike all the instruments in a symphony orchestra that all have to play their parts together. As with some modern orchestras, there is no conductor. And if parts of the orchestra don't perform properly, the whole thing falls apart.

Everything we have discussed so far is about the different ways that cells sense when things are not right and what they do to correct that. This is an amazingly complicated web of interactions, which is itself controlled by yet more proteins. If the control proteins themselves become defective, the problems are amplified. That is just what happens as we age.

We began this chapter with the terrible scourge of Alzheimer's disease. The disease, which is increasingly a dread of old age, turns out to be related to a curious group of diseases whose cause was uncovered in a most unexpected way. The key person to unravel its mystery was Carleton Gajdusek, a scientist with the unique and unfortunate distinction of being both a Nobel Prize winner and a convicted child molester.

After earning his medical degree from Harvard, Gajdusek was serving a fellowship in Boston when he was drafted into the army. He ended up in the Korean War, where he showed that a fever that was killing American soldiers was spread by migrating birds. On the strength of this, he was offered a job with the US government's Center for Disease Control, but chose instead to work with the famous immunologist MacFarlane Burnet in Melbourne, Australia. Burnet sent him to Port Moresby, New Guinea, to set up part of a multinational study on child development, behavior, and disease. It could not have been easy carrying out fieldwork in such a remote area, far away from any modern research laboratory, but Gajdusek was an unusual character. Burnet once described him as someone who "had an intelligence quotient up in the 180s and the emotional immaturity of a 15-year-old," adding candidly that his protégé was completely self-centered, thick-skinned, and inconsiderate. At the same time, said Burnet, the young man from the United States would not let the threat of danger, physical hardship—or other people's feelings—interfere in the least with what he wanted to do.

While in Port Moresby, Gajdusek heard about a mysterious illness called kuru and set out for the Eastern Highlands Province, about 200 miles away, where the disease was prevalent among the native Fore tribe. Patients with the disease showed no symptoms of fever or inflammation but died of a progressive brain disease that caused tremors and highly abnormal behavior such as uncontrolled fits of laughter. Two anthropologists, Shirley Lindenbaum and Robert Glasse, observed that women and children, but not adult men, ate the entire bodies of deceased family members, even the bones. This was a recent practice among the Fore, and by collecting detailed evidence of cannibal feasts which could be matched with the subsequent appearance of the disease in participants, they concluded that this practice of cannibalism may have had something to do with transmission of the disease. Gajdusek and a colleague named Vincent Zigas had observed that one of the practices of the tribe was to cook and eat the brains of deceased family members following funerals. So Gajdusek suspected that something in the diseased brain was transmitting the disease to the people who ate it. Following up on this hunch, he was able to show that you could transmit kuru to chimpanzees by injecting their brains with extracts from the brains of diseased patients.

The autopsied brains of the Fore tribe, when examined under a microscope, were full of holes, like a sponge. Kuru is one of many brain diseases with this pattern, called spongiform encephalopathies, including a variant form of Creutzfeldt-Jakob disease. (*Variant* refers to the transmissible rather than inherited form of a disease.) About 10 percent of all cases are inherited, and just as he had done for kuru, Gajdusek was able to show that brain extracts from infected patients could transmit the disease to chimpanzees. The idea that a disease could be inherited in some instances but also *transmitted* like an infection in other cases was unprecedented. Gajdusek was awarded a Nobel Prize in 1976.

Unfortunately, the end of Gajdusek's career was not so glorious. Over the course of many years, he brought back more than fifty children to the United States from New Guinea and Micronesia, and acted as their guardian. In the 1990s, in response to a tip-off from a member of his lab, the FBI began to investigate the scientist. The bureau persuaded one of the boys to tape a phone conversation in which Gajdusek admitted that he and the boy had sexual contact. In a plea bargain that would be unthinkable today, he served a year in jail in 1997 and then left the United States as soon as he was released to spend the rest of his life in Europe. During his self-imposed exile, he stayed active scientifically and was affiliated with several universities. He showed no remorse for his behavior, dismissing his treatment as American prudishness. Many of the boys continued to have contact with him, some adopting his name and even naming their own children after him. In 2008 he died in a hotel room in Tromso, Norway, where he was a frequent visitor to the university there.

Gajdusek's concept of transmissibility had a huge impact on our thinking about this class of diseases. Mad cow disease (bovine spongiform encephalopathy) afflicted cows in Britain, notably in the 1980s, as a result of cows being fed the remnants of infected animals. Around this time, more than a hundred people died of Creutzfeldt-Jakob disease. Scientists began to suspect that this was because they had eaten meat from diseased cows. The connection with eating infected beef was then not universally accepted, and John Gummer, a UK government minister, famously encouraged his four-year-old daughter, Cordelia, to eat a hamburger on television, declaring British beef to be completely safe. (The girl did not get sick.) Nevertheless, many countries prudently banned the importation of British beef and lifted it only after several million cows had been slaughtered and farming practices had been changed.

Although the transmissibility of these diseases was established, it was not clear exactly how they spread. Ever since the nineteenth and early twentieth centuries, it has become a firm dogma that every infectious disease is transmitted by living organisms that can multiply in the host, whether they are parasites or microbial organisms such as bacteria, fungi, or viruses. In the early 1980s Stanley Prusiner, an American neurologist at the University of California, San Francisco, began trying to isolate the infectious agent for scrapie, a spongiform encephalopathy of sheep and goats. The brain extracts that transmit scrapie remained infectious even after they were sterilized using standard methods such as heat, so the prevailing view was that the infectious agent was a virus that was resistant to inactivation and had a long incubation time. When Prusiner gradually isolated the infectious agent, it turned out to be a protein—a notion that was greeted with a chorus of skepticism. After all, unlike bacteria or viruses, proteins could not multiply, so how could they possibly cause an infection that spread from one animal to another?

Over the next several years, Prusiner identified the protein and showed that although it was a normal component of brains, its shape in a scrapie-infected brain was abnormal. Prusiner called the protein a prion and proposed there were two forms: a normal version and a scrapie version. Like an evil character who corrupts all the good people around him, this aberrant, misfolded, scrapie version of the protein acts as a mold, or template, and induces each normal prion protein it encounters to switch to the misfolded version. The result is that the misfolded form spreads like an infection throughout the cell and across cells throughout the tissue, bringing about disease.

At first glance, the only commonality between diseases such as kuru or scrapie and Alzheimer's is that they are lethal brain diseases, but as we shall see, the similarity runs deeper. Dr. Alois Alzheimer himself autopsied the brains of deceased patients and discovered deposits of plaques outside cells as well as tangles of fibrils inside some nerve cells. It wasn't initially clear whether the formation of these deposits was a cause of the disease or a symptom.

In 1984, scientists identified that the major component of the plaques was a protein called amyloid-beta, which itself is produced by trimming a much larger amyloid precursor protein, or APP. Alzheimer's is normally a disease of old age and not necessarily inherited, but some patients with inherited forms develop the disease earlier in life. They turn out to have mutations in the APP gene. Scientists have also identified the enzymes that trim the APP to the mature amyloid-beta and, in a nod to their involvement in causing senility, called them presenilins. Mutations in these proteins also led to familial Alzheimer's disease. The case that the disease was caused by accumulating either too much or incorrectly processed amyloid-beta protein seemed overwhelming. Much of the research community then focused on the details of what caused the plaques to develop and how they could be prevented.

However, in science, things are often never quite so straight-forward. For one thing, the plaques typically develop outside nerve cells, so why are they killing them? Another curious feature is that other tissues—for example, blood vessels—also contain amyloid-beta deposits, but it is the diseased brain that kills people. A feature of the disease that was ignored earlier on is that inside some neurons of patients, there are filaments made of a different protein called tau. Perhaps these tau filaments were the cause of the disease?

Although scientists were skeptical at first, evidence incriminating tau also began to mount when three groups found independently that patients with an inherited form of dementia related to Parkinson's disease had mutations in the tau gene. Also, it was not hard to imagine how tau could cause disease. The tau filaments could block the narrow axons and dendrites that connect neurons, and, not surprisingly, it is these connections that are the first to go, causing cognitive impairment.

Recently, scientists have found that the filaments characteristic of diseased brains are not just random clumps of unfolded proteins. Rather, the aberrant molecules come together to form filaments that are distinct for each type of dementia. Studies show consistently that the tangles we see in diseased brains actually have very well-defined structures, each of which is a hallmark of a particular disease. This is something we did not know even a few years ago.

Therefore, as things stand, we have very compelling evidence that amyloid-beta, tau, and other filaments are implicated in disease. One problem is that nobody really understands what these proteins are doing normally. We do know that if you delete the genes for them in mice, the animals exhibit some abnormalities, but they don't develop plaques or Alzheimer's disease. This means that the reason amyloid-beta or tau causes disease is not because it has ceased to function normally. Rather, it is because the unfolded forms can give rise to filaments that spread throughout the brain.

Alzheimer's and prion diseases are both caused by aberrant forms of proteins that come together to form tangles or plaques. In prion diseases, the prion form assumes a different shape from the normal form, and spreads because it switches the normal version into the prion form when it comes into contact with it. There is a growing feeling that exactly the same thing happens in Alzheimer's and other neurodegenerative diseases: an abnormal, unfolded form can seed the formation of filaments, which then spread throughout the brain. Injecting brain extracts from Alzheimer's disease patients into mice stimulates the premature formation of plaques or tangles. But, unlike prion diseases such as kuru and bovine spongiform encephalopathy, nobody has demonstrated that Alzheimer's, Parkinson's, or similar diseases are actually infectious. That could be because we don't eat the brains of patients with dementia or inject extracts of their diseased brains into our own.

What causes Alzheimer's disease is a burning question because that holds the key to preventing it. The answer depends on how you define cause. The immediate cause may well be the formation of tau or amyloid-beta filaments in the brain. However, an earlier and root cause is the cell's inability to manage the excess of unfolded proteins that aggregate to form these filaments in the first place. This in turn is caused by damage to our control systems: the quality control and recycling machinery of the cell that we discussed earlier in the chapter. And that damage to our control systems is a result of aging.

So you could say it all boils down to our living long enough for the damage to occur. It is particularly ironic that one of the consequences of our increased life expectancy over the last century is the greater likelihood of spending our final years with the terrible effects of diseases such as Alzheimer's.

Can anything be done about it? The difficult truth is that there are still no effective treatments for these dementias, despite several decades of work. Just as cancer is so hard to treat because it is our own cells that have gone out of control, Alzheimer's is caused by our own proteins misbehaving. And just as with cancer, there may be both genetic factors and chemicals or infectious agents that accelerate the process. This creates a fundamental difficulty for treatments. Very recently, therapies based on antibodies that bind to the amyloid-beta protein were shown to halt cognitive decline by about 25 percent after eighteen months. They were most effective at slowing the progression of the disease if treated early, and in patients that had only a modest level of tau aggregates. They carried a serious risk of side effects, including seizures and bleeding in the brain. However, they did demonstrate that targeting beta-amyloid showed some clinical effect, and against the bleak backdrop of having next to nothing to offer Alzheimer's patients, even an expensive and complicated treatment with a relatively modest gain was heralded as a huge breakthrough.

All the recent breakthroughs in our understanding the basis of the disease offer some hope, however. Now that we know that the filaments are not random but consist of very specific contacts to form their structure, perhaps drugs can be developed to prevent their formation. Others are attempting to inhibit the production of the protein itself. And scientists are busy at work on the ultimate causes as well, including how to modify aging cells so they can handle aberrant proteins as effectively as younger cells do. We also need to identify suitable biomarkers that are an early warning of incipient disease. As we learn much more about the underlying biology involved, we can be hopeful that we will find more ways to prevent the disease in the first place, and diagnose it early and treat it when it occurs.

7.

LESS IS MORE

The India in which I grew up is a land of many religions, and there never seemed to be a time when one or another group wasn't fasting. Hindus fasted before certain religious occasions—or if they were strict, every week. Muslims fasted from dawn to dusk for the entire month of Ramadan, not drinking a drop of water even when the holiday fell amid the long, hot summer days of the subcontinent. Christians fasted during Lent. And fasting was not only a religious imperative. Nearly all cultures considered fasting, and moderation in general, a key to a long and healthy life, and gluttony to be a vice.

For much of our existence as a species, we were hunter-gatherers, feasting occasionally between prolonged periods of involuntary fasting. Perhaps our metabolism evolved to adapt to that lifestyle. It is different today, especially in the rich countries of the West. Like millions of others, I gained an inordinate amount of weight during the early days of the Covid-19 pandemic, when most people were stuck at home, and food was only as far away as the refrigerator. Indeed, today we face a widespread epidemic of obesity, which is linked not only to cardiovascular disease and type 2 diabetes but also to certain cancers and even Alzheimer's disease. It is also a major risk factor in infections: Covid-19 patients who were obese were far more likely to die from the virus. Clearly it has far-reaching consequences, both for ill health in old age and our likelihood of dying from those disorders.

The reasons for the rise in obesity in recent times are complex. One popular theory is that throughout most of our history, food was scarce and sporadic, and those who had "thrifty genes" that could store fat more efficiently could better survive times of scarcity. Now, in a time of plenty, those very genes efficiently keep storing away all the **excess** fat we eat and cause obesity. This idea was so prevalent that it became a truism, but it is now being questioned. Even today, less than half the population in the United States is obese. John Speakman, who has studied the relationship between energy intake and weight in organisms, has argued convincingly that it is simply that the population had a lot of genetic variability in how efficiently they could store fat, a variability he calls "drifty genes." When food was generally scarce, even those individuals who might be prone to becoming obese rarely were. But now, an abundance of calorie-rich food has driven a rise in obesity, especially in the portion of people who have inherited genes that in previous eras would not have caused any harm. Also, historically there was no reason for us to have evolved to be abstemious.

Regardless of the reasons for the rise in obesity, nobody doubts that moderation and maintaining a healthy weight are recipes for good health. Clearly, overeating is bad for your health, but is the converse also true? Would stringently restricting our diet to less than what we eat normally actually make us live much longer? The first studies to test this, carried out in 1917, were not taken seriously, perhaps because for most of our existence as a species, being undernourished was a much greater threat to life than overeating. Nevertheless, the idea persisted, and later studies showed that rats fed a calorie-restricted diet lived longer and were healthier than those allowed to eat without limit.

During caloric restriction, or CR, an animal is fed 30–50

percent fewer calories than it would consume if it ate as much as it liked (ad libitum), while making sure that it consumes enough essential nutrients to not become malnourished. In rodents and other species, animals on CR lived 20–50 percent longer, as judged by both average life span and maximum life span. Moreover, they appeared to have delayed the onset of several diseases of aging, including diabetes, cardiovascular disease, cognitive decline, and cancer.

Mice are small, however, with short life spans. What about animals more similar to us? In 2009 a long-term study from the University of Wisconsin found that rhesus monkeys lived longer and were healthier and more youthful when subjected to caloric restriction. But this was contradicted only a few years later by a twenty-five-year study at the National Institute on Aging (NIA). The Wisconsin diet was richer and had a higher sugar content, so perhaps eating a healthy diet rather than fewer calories might have made the difference. The NIA control animals were not allowed to eat ad libitum but were fed an apportioned amount to prevent obesity. More than 40 percent of the Wisconsin control group developed diabetes, while only 12.5 percent of the NIA control group did. In tandem, the studies suggest that for animals already on a healthy diet and not overweight, further caloric restriction has little additional effect on longevity. Interestingly, all the animals in both groups, even the CR animals, weighed more than animals found in the wild, suggesting that even the restricted diet provided more food than they would eat naturally.

Experimenting with monkeys is hard enough. They can live between twenty-five and forty years, and the studies from NIH and Wisconsin have gone on for over two decades and already cost millions of dollars. Conducting similar studies with humans—who live more than twice as long and whose dietary intake is much harder to track—seems out of the question. Any evidence for the effect of CR on human longevity is purely anecdotal at this point, but that hasn't stopped individuals from experimenting on themselves and even writing books to tout their lifestyles.

There have also been persistent claims that fasting is beneficial for health beyond simply reducing the overall intake of food. There is 5:2 fasting, whose adherents eat as little as 500–600 calories per day twice a week but eat normally on the other five. Another method advocates eating all your food in a window of a few hours each day. Recently, scientists examined the effects not just of CR and intermittent fasting in mice but also of aligning feeding times to their daily biological rhythms. They concluded that matching feeding times to our biological circadian rhythm greatly improved the benefit of intermittent fasting. This might seem like the home run the field wanted, but, as the accompanying commentary points out, much of the additional benefit may have nothing to do with the time of feeding as such. Rather, if you allowed mice to eat only during the day—when they would normally be asleep—they were faced with the unenviable choice between starving and not sleeping. The test animals chose to disrupt their sleep. Even if you distributed the restricted diet throughout the twenty-four-hour period, the mice would not get enough to eat when they were awake and would choose to disrupt their sleep to get the rest.

I know what a wreck I am when I am sleep deprived. As I get older, my problems with jet lag are getting worse, and I am barely able to function right after I show up on some other continent. So I am always struck by how sleep, which is so intimately related to our health, is ignored by scientists in other fields. We think of sleep as something that is connected with our brains and especially our eyes and vision. But as Matthew Walker explains so well in his book *Why We Sleep*, you don't need a brain or even a nervous system to sleep. In fact, sleep is ancient and highly conserved across the entire kingdom of life. Even single-celled life forms follow a daily rhythm that is related to sleep. Considering that sleep can be perilous—animals are vulnerable to attack when they are asleep—it must have huge biological benefits for it to persist through evolution. The consequences of sleep on our health are profound and widespread. In particular, sleep deprivation increases the risk of many diseases of aging, including cardiovascular disease, obesity, cancer, and Alzheimer's disease. According to a recent study, one of the ways that a lack of sleep accelerates aging and death is by altering repair mechanisms that prevent the buildup of damage to our cells.

But going back to the study matching feeding times with when mice are awake, although it did not explicitly monitor the sleep patterns of the mice, the researchers suggest that as long as you don't deliberately disrupt sleep, CR has a significant positive effect on both health and longevity. Over the decades, study after study have confirmed the benefits of CR over an ad libitum diet in multiple species.

If all this seems too good to be true, it might be. In one study, the effects of CR varied greatly depending on the strain and sex of the mice; in fact, in a majority of the test animals, CR actually *reduced* life span. Indeed, one of the pioneers of the aging research field, Leonard Hayflick, expressed skepticism that dietary restriction had any effect on aging. He felt that animals on an ad libitum diet were overfed, and unhealthy as a result, and caloric restriction simply brought their diets closer to conditions in the wild. Moreover, when scientists look outside typical lab conditions to animals in the wild, the link between eating less and living longer becomes much more tenuous.

Nevertheless, in multiple laboratory studies, at least compared to an ad libitum diet, CR appears to be beneficial not only in rats and mice but also in diverse organisms ranging from worms, to flies, to even the humble unicellular yeast. Most scientists working on aging agree that dietary restriction can extend both healthy life and overall life span in mice and also leads to reductions in cancer, diabetes, and overall mortality in humans. On a more granular level, limiting protein intake or even just reducing consumption of specific amino acids such as methionine and tryptophan (both of which are essential in our diets because our bodies don't produce them) can confer at least some of the advantages of overall dietary restriction.

It might seem counterintuitive that eating the bare minimum to avoid malnutrition would be good for you. In fact, the results of CR may be yet another example of the evolutionary theories of aging. Consuming lots of calories allows us to grow fast and reproduce more at a younger age, but it comes at the cost of accelerated disease and death later on.

So why aren't we all on CR diets? For the same reason that rich countries face an epidemic of obesity: we now live in a time of plentiful food, and we have not evolved to be abstemious. Moreover, caloric restriction is not without its drawbacks. It can slow down wound healing, make you more prone to infection, and cause you to lose muscle mass, all serious problems in old age. Among its other reported downsides are a feeling of being cold due to reduced body temperature, and a loss of libido. And, of course, a side effect that to most readers will seem blindingly obvious: people on calorically restricted diets feel perpetually hungry. In fact, animals on CR diets all revert to eating as much as possible when permitted.

The anti-aging industry would love to produce a pill that can mimic the effects of CR without our having to forego the ice cream and blueberry pie. For that to happen, we need to understand exactly what caloric restriction does to our metabolism. It's a story full of unusual twists and turns and the discovery of some completely new processes in our cells.

IN 1964 A GROUP OF Canadian scientists set out on a voyage to Easter Island, a remote spot in the South Pacific that is about 1,500 miles away from its nearest inhabited neighbor. Their goal was to study the common diseases of the island's Indigenous people, who had little contact with the outside world. In particular, they wanted to know why the islanders did not develop tetanus, even though they walked around barefoot. The researchers collected sixty-seven soil samples from different parts of the island. Only one of them had any tetanus spores, which are typically more common in cultivated soil that has less diversity of microbes than virgin soil does. Nothing further might have come out of this expedition had not one of the scientists given the soil samples to the Montreal lab of Ayerst Laboratories, a pharmaceutical manufacturer. The company was looking for medicinal compounds produced by bacteria. By then, it was well known that soil bacteria, notably the genus *Streptomyces*, produced all kinds of interesting chemicals, including many of the most useful antibiotics today. Part of the reason they produce them is thought to be biological warfare among soil microbes, where some species make compounds that are toxic to others.

To identify anything useful from an unknown bacterium in a soil sample, you first have to isolate it and coax it to grow in the lab. Then you need to analyze the hundreds or thousands of compounds that it makes and screen them for useful properties. Through this painstaking venture, the Ayerst scientists found that one of the vials contained a bacterium, *Streptomyces hygroscopicus*, that made a compound that could inhibit the growth of fungi. Because fungi are more similar to us than bacteria are, it is hard to find compounds that will treat fungal infections without also harming our own cells. So it seemed worthwhile to follow up on their initial observation. It took Ayerst two years to isolate the active compound, which the company named *rapamycin* after Rapa Nui, the Indigenous name for Easter Island.

The scientists soon discovered that rapamycin had another, potentially much more useful property. It was a potent immuno-suppressant and stopped cells from multiplying. Suren Sehgal, a scientist at Ayerst, sent off some of the compound to the US National Cancer Institute. Researchers there found the drug to be effective against solid tumors, which are ordinarily difficult to treat. Despite these promising early results, work on rapamycin ground to a halt when Ayerst closed its Montreal lab and relocated the staff to a new research facility in Princeton, New Jersey, in 1982.

Sehgal, however, was convinced that rapamycin was going to be useful. Just before moving to the States, he grew a large batch of *Streptomyces hygroscopicus* and packed it into vials. At home, he stored them in his freezer next to a carton of ice cream, with a label cautioning, "Don't Eat!" The vials remained there for years. In 1987 Ayerst merged with Wyeth Laboratories, and Sehgal persuaded his new boss there to pursue rapamycin. He was given the go-ahead to look at its immunosuppressive properties, which could be useful to prevent transplant rejection. Eventually rapamycin was approved as an immunosuppressant for transplant rejection, but nobody had any real idea of how it worked. How could it inhibit the growth of fungi, prevent cells from multiplying, and be an immunosuppressant, all at once?

Here our story shifts to Basel, Switzerland, where two Americans and an Indian chanced upon an unexpected breakthrough. One of the Americans, Michael Hall, had an unusually international childhood: he was born in Puerto Rico to a father who worked for a multinational company and a mother who had a degree in

Spanish. They both liked Latin American culture and decided to make their home in South America, where Hall grew up, first in Peru and then in Venezuela. When he was thirteen, his parents decided he needed a rigorous American education; Hall was suddenly ejected from his carefree life wearing T-shirts, shorts, and sandals in warm and sunny Venezuela, and dropped into a boarding school in the freezing winters of Massachusetts. From there he attended the University of North Carolina, intending to major in art but eventually settling on zoology, with the intention of going to medical school. An undergraduate research project whetted his appetite for science, and Hall went on to earn a PhD from Harvard and then put in time pursuing postdoctoral research at the University of California, San Francisco. In between, he spent almost a year at the famous Pasteur Institute in Paris, where he met Sabine, the Frenchwoman who would become his wife. Thus, unlike many American scientists who see leaving the United States as equivalent to falling off the map, Hall cast a broad net in the job search that followed his postdoc. He had not originally thought of moving to Switzerland, but when he interviewed for a starting faculty job at the Biozentrum at the University of Basel, he fell in love with the institute and the city.

Shortly after he started his lab in Basel, Hall was joined by another young American, Joe Heitman, who was in an MD-PhD program that combined medical studies at Cornell Medical School with research at Rockefeller University. After his PhD research, rather than go back immediately and finish his medical degree, Heitman decided to do some postdoctoral research, partly because his wife would be starting her own postdoctoral work in Lausanne, Switzerland. Looking for suitable labs in the vicinity, he identified Hall as someone he wanted to work with. His initial project there turned out to be frustrating, however, and Heitman briefly considered going back to medical school, when he read a scientific paper describing mutants of a mold, *Neurospora*, that were resistant to the immunosuppressive drug cyclosporine. He approached Hall with the idea of studying immunosuppressants using yeast.

By sheer chance, Heitman could not have found a more receptive mentor. It turned out that cyclosporine was a blockbuster drug for Sandoz, the pharmaceutical company located right in Basel, and Hall had already begun working with a scientist there who was interested in how it and other immunosuppressants worked. That scientist, Rao Movva, who grew up in a small village in India, had already enjoyed quite a bit of success in using yeast to understand the mechanism of cyclosporine, and he was keen to study rapamycin, which was still being developed for use in patients.

To most in the field, this must have seemed a crazy idea. What could yeast—a unicellular organism that doesn't *have* an immune system—teach them about immunosuppressive drugs and human beings? But Hall points out that these compounds were produced as part of biological warfare among soil microbes, so, really, yeast was their natural target; it is administering them to humans that is actually unnatural. As soon as Heitman had expressed interest in the problem, Hall put him in touch with Movva. This was a huge advantage, because at a large pharmaceutical company such as Sandoz, Movva had the resources to produce enough rapamycin. One day he came into Hall's lab with a small vial and told Heitman, "Okay, this is the world's supply of rapamycin. Think very carefully about the next experiments you're going to do. Don't blow it, because this is all we have."

The gamble paid off. The trio looked for mutant strains of yeast that would grow even in the presence of rapamycin, and their experiments revealed that many of the mutations occurred on two closely related new genes that coded for some of the largest proteins in yeast. Names of genes and proteins from yeast typically consist of a three-letter acronym that makes little sense to those outside a particular field. In this case, from a long list of possibilities, they chose TOR1 and TOR2, to denote "target of rapamycin." The names held additional appeal for Heitman because he lived near one of the picturesque medieval gates of Basel, and the German word for gate is *Tor.*

This was a big breakthrough. Rapamycin's immunosuppressive activity was thought to derive from its ability to inhibit cell growth. The compound also arrests yeast growth, however, so identifying its protein targets would enable scientists to understand exactly how. The mutants identified two genes, but without cloning and sequencing them, nothing was known about the proteins they coded for, let alone what they did.

At this point, the problem almost fizzled out in Hall's lab. Heitman stayed as long as he could, but he had to return to New York to finish his medical studies. At the time, although it was acknowledged that rapamycin was a potentially important immunosuppressive drug, nobody had any idea of how important their discovery would turn out to be. Meanwhile, Heitman's mutants were sitting in the lab freezer until a new student was frustrated when her original project was not working. She, along with another student and others in the lab, used the mutants to clone and sequence the TOR1 and TOR2 genes. In those days, sequencing had to be done manually. What's more, this was no trivial project, because they were both among the largest genes in yeast, and were similar but not identical. One of them was lethal when deleted, proving that it was essential in order for yeast to survive, while the other was not.

Understanding the mechanism of an immunosuppressive drug that was also a potential anticancer drug was of great medical importance, so while Hall and his colleagues carried on their work, they were participants in an intense race to discover the target of rapamycin. Three groups in the United States directly purified the protein target of rapamycin in mammals. It turned out to be the mammalian counterpart of the genes that Hall and his colleagues had identified. Now, scientists can be fiercely competitive and don't like to come in second place. It's a bit like leading the *second* expedition to climb Mount Everest or being the *second* pair of astronauts to walk on the moon—you just don't get the same level of recognition. In the case of the two genes, prickly egos and difficulty accepting one's also-ran status led to a profusion of names in the field, sowing confusion.

The US research groups realized that they had discovered the mammalian version of essentially the same protein that Hall and his colleagues had identified already. Nevertheless, some of them gave it entirely different names. Eventually they all agreed to christen it mTOR, with the *m* denoting "mammalian," to distinguish their findings from the yeast TOR. When the same protein was identified in a variety of organisms, including flies, fish, and worms, things began to get a little silly, with scientists studying zebrafish calling their version zTOR or DrTOR (the scientific name for zebrafish is *Danio rerio*). Eventually everyone settled on mTOR for all species—except, paradoxically, the original yeast!—with the *m* now standing for *mechanistic*, which makes no sense at all, since it implies that there is also some other target of rapamycin that is nonmechanistic (whatever that means). Why they didn't simply revert to the original TOR remains a mystery to me. For consistency, and in deference to the original discoverers, I will refer to the molecule as TOR, but if you read elsewhere about TOR with a small letter before it, it is basically referring to the same protein.

From the start, it was known that rapamycin would prevent cultures of cells from growing, but it wasn't clear how. Did it limit the number of cells or the average size of each cell? At first, Hall thought that rapamycin would simply stop cells from dividing, but after pushback from a famous expert in that field, he realized that TOR actually controlled cell growth by activating the synthesis of proteins in the cell when nutrients are available. Among other things, Hall and his colleagues showed that in the presence of rapamycin, or mutants of TOR, cells would appear starved and stop growing even when plenty of nutrients were available.

Biologists have known for a very long time that the size and shape of cells is highly controlled. Cell size varies not only in different species but also in different tissues and organs. For example, an egg cell is about thirty times the diameter of the head of a sperm cell, and neurons can have protrusions, the nerve axons, as long as three feet. How cell size and shape are controlled is still a very active area of research. But the general belief was that cells would simply keep growing and dividing as long as you provided them nutrients—unless, that is, they received specific signals to stop growing. Hall's experiments turned this dogma around. Cell growth, they suggested, was not passive; rather, TOR had to actively stimulate it, by sensing when nutrients were present.

It is a bit like the difference between an old steam locomotive and a gasoline-powered car. Once a locomotive gets going, as long as it has plenty of burning coal in the furnace and water in the boiler, it will keep rumbling down the track unless you take action to stop it. But a car, even with a full tank of gas, requires a foot on the accelerator in order for the vehicle to remain in motion; you have to actively do something to use the fuel. TOR is the driver that presses on the gas pedal to ensure that available nutrients are used to drive cell growth.

Hall's conclusions represented a paradigm shift in our understanding of how cells grow and ran counter to decades of understanding. His paper was rejected seven times before it found a home in the journal *Molecular Biology of the Cell* in 1996. Around the same time, Hall also collaborated with Nahum Sonenberg, the same scientist we encountered in chapter 6 for his studies on the integrated stress response, and who is best known for his work on how ribosomes initiate; in other words, how they find the beginning of the coding sequence on mRNA and start reading it to make proteins. They found that without TOR actively making it possible, cells could not begin the process of translating mRNA to produce proteins, and would stop growing.

The initial discoveries by Hall and the other groups opened up the floodgates. Since then, TOR has become one of the most studied molecules in biology with about 7,500 research articles in 2021 alone. There is no question that finding out how rapamycin was immunosuppressive was important. But not even the brilliant scientists first working on it could have imagined that they would later uncover one of the oldest and most important metabolic hubs of the cell. In metabolism, proteins seldom act in isolation; they influence the actions of other proteins. If you think of such proteins as nodes that connect to one another—picture an airline map of its routes—TOR would be a major hub like London, Chicago, or Singapore, making direct connections to a large number of cities all over the world.

How could one protein have such widespread effects on the cell, and how exactly was it linked to caloric restriction? Ever since Michael Hall and his colleagues sequenced the two TOR genes, we have known that TOR is a member of a family of proteins called kinases. These enzymes often act as switches by adding phosphate groups to other proteins, which then act as tags or flags to turn them on or off. (The act of adding phosphate groups is called phosphorylation, and the proteins with the added phosphates are described as phosphorylated.) Sometimes kinases activate other kinases, which in turn activate other enzymes. You can think of kinases as part of a huge relay system, where many different proteins in a large network are turned on or off in response to some cue in the environment or the state of the cell. A map of all the proteins involved in activating or being activated by TOR is enormously complicated. So it is not surprising that by responding to many different environmental cues and then switching on or off many different targets, TOR has such widespread effects within the cell. Some of these environmental cues are not sensed directly by TOR but by other proteins, which in turn activate TOR.

TOR is not a protein chain that functions all by itself. It is part of two larger complexes called TORC1 and TORC2. Much more is known about TORC1, which is activated by proteins that sense the level of nutrients such as individual amino acids and hormones, including those that stimulate growth, known as growth factors. It is also affected by energy levels in the cell. If conditions are right, TORC1 promotes the synthesis not only of proteins but also nucleotides, which are the building blocks of DNA and RNA, and also lipids, which make up the membranes of all cells and organelles.

An important function of TOR is that when nutrients are available and the cell is not stressed, it inhibits autophagy, which, as you learned in chapter 6, is the process by which damaged or unneeded components of the cell are taken to the lysosome to be destroyed and recycled. This makes sense because these are exactly the conditions in which you want to stimulate cell growth and proliferation, not the opposite.

We can now see how TOR is connected to caloric restriction. Under CR, there are fewer nutrients around, and TOR, recognizing that, can switch off protein synthesis and other growth pathways, and also green-light autophagy. We have already seen how important both controlling protein synthesis and clearing defective proteins and other structures through autophagy are to keep the cell working optimally, and to aging in general.

But what if we didn't need caloric restriction to reap its

benefits—if we could inhibit a normal TOR and mimic its effects, with no change to the human diet? TOR was discovered precisely because it was the target of rapamycin. Might rapamycin be the long-sought pill that could imitate CR without our having to cut down on how much we eat?

It turns out that both a defective TOR and inhibiting TOR with rapamycin can enhance health as well as longevity in a range of organisms, from the simple yeast, to flies, to worms, and to mice. Strikingly, even short courses of rapamycin, or initiating treatment relatively late in the life of mice (equivalent to age sixty in men and women), conferred significant improvements in both health and life span. Rapamycin also delayed the onset of Huntington's disease in a specially engineered strain of mice, presumably because it increased autophagy and prevented the accumulation of misfolded proteins. This shows that rapamycin not only improves longevity, but may also keep the mice healthier. In fact, the two may be closely related—perhaps the mice in these experiments live longer precisely because they are protected against various disorders of aging.

Though rapamycin is an immunosuppressive drug, it also, counterintuitively, improves some aspects of our immune response. There are two important components of our immune system: one is B cells, a type of white blood cell that churns out antibodies for identifying and then binding to the surfaces of bacteria, viruses, and other foreign invaders, or antigens, so that other foot soldiers in the body's self-defense corps can race to the crime scene and finish off the culprit. The other is T cells, another type of white blood cell: helper T cells stimulate B cells to manufacture antibodies, while killer T cells, as their name implies, recognize and destroy cells that have been infected by a pathogen. While rapamycin inhibits those parts of the immune system responsible for rejecting grafts of tissue from a donor (such as kidney, bone marrow, or liver transplantation) and triggering inflammation in general, it actually increases the functional quality of certain helper T cells, thus potentially improving a person's response to vaccines. Another study, from 2009, showed that administering rapamycin in mice rejuvenates aging hematopoietic stem cells, the precursors of the cells of the immune system, and boosts the body's response to the influenza vaccination.

These results generated a great deal of excitement about rapamycin in the anti-aging community, but before we charge ahead with an immunosuppressive drug as a long-term panacea against aging, a note of caution is warranted. As one might expect, numerous studies have warned that long-term rapamycin use increases the risk of infection, such as with cancer patients. In fact, in that seemingly encouraging 2009 mouse study, treatment with rapamycin had to be paused for two weeks prior to administering the vaccine, the authors acknowledged, to "avoid the possible suppression of the immune response by rapamycin." It makes one wonder whether the results would have been as promising without the pause to clear away the rapamycin.

Moreover, it is possible that some of the effects of rapamycin and TOR inhibitors are due to a general reduction of inflammation. Yet other research contends that optimal health calls for a fine balance between excessive inflammation and heightened susceptibility to infection. In a recent study, scientists show that TOR inhibitors dramatically increase the susceptibility of zebrafish to pathogenic mycobacteria closely related to the bacteria that cause TB in humans, and point out that this "warrants caution in their use as anti-aging or immune boosting therapies in the many areas of the world with a high burden of TB."

Still, rapamycin's draw as a potential wonder drug endures. In some quarters, the excitement has overtaken the data: one prominent aging researcher told me that he knew several scientists who were quietly self-medicating with rapamycin. I asked Michael Hall what he thought about using an immunosuppressive drug to combat aging, and he replied, "I suppose the rapamycin advocates are following Paracelsus's adage that the poison is in the dose." He was alluding to the Renaissance Era Swiss physician who defended his use of substances that he believed were medicinal even though they were toxic at higher doses. In fact, most drugs, even relatively safe ones such as aspirin, can be toxic if the dose is high enough. It may well be that low or intermittent doses of rapamycin or other TOR inhibitors can confer most of their benefits without serious risks. But we need long-term studies on their safety and efficacy before they can be used to target aging in humans.

A problem with laboratory animals, including mice, is that they are kept in a highly protected and relatively sterile environment that does not mimic real-life conditions. To address this, Matt Kaeberlein at the University of Washington in Seattle is leading a nationwide US consortium to study the health and longevity of domestic dogs. Canines not only vary greatly in size but also live in environments as diverse as their owners', so this is a way to conduct controlled studies in a natural setting outside of a laboratory environment. The consortium will analyze various aspects of dogs' metabolism, including their microbiome and the differences between how large dogs age compared to small dogs. It will also carry out a randomized study on the effect of rapamycin in large middle-aged dogs. Experiments like these will go a long way to establishing whether rapamycin will turn out to be useful for general health in old age.

It is curious that using rapamycin to shut down a major pathway in the cell could actually be beneficial. As is often the case, the answer to this paradox lies in the evolutionary theories of aging discussed earlier. In a 2009 paper published in the journal *Aging*, Michael Hall, of the University of Basel, and the Russian-born evolutionary biologist Mikhail Blagosklonny suggest an explanation: TOR promotes cell growth, which is essential in early life. Later, however, it is unable to switch itself off even when the growth it drives becomes excessive, leading to cell deterioration and the onset of age-related diseases. They go on to suggest that while these pathways that cause aging cannot be completely switched off by a mutation (because that would be harmful or even lethal early in life), perhaps they can be inhibited by drugs such as rapamycin years later, when an uninhibited TOR becomes a problem after individuals have reached middle age.

This chapter began with how the age-old idea of fasting as a beneficial practice gained credence with scientific studies on caloric restriction. However, the journey to discover a potential drug that could replicate the advantages of restricting calories without requiring unwavering self-control is nothing short of extraordinary. It began with a completely open-ended fishing expedition by Canadian scientists to find something interesting in the soil of the remote island of Rapa Nui. Just one of many soil samples they collected had a bacterium that produced a promising compound, and that nearly died in a scientist's freezer as he moved from one country to another. The baton was taken up years later by two Americans and an Indian working in Switzerland. None of the scientists involved had any idea that they would be revealing one of the cell's most important pathways with connections to both cancer and aging. This is often how science works: people follow their curiosity, and one thing leads to another. It is a story of persistence, insight, brilliance, and vision, but also chance encounters and sheer luck. If this strange journey ends up unlocking a key to protecting us from the relentless onslaught of old age, it would indeed be a scientific miracle.

8.

LESSONS FROM A

LOWLY WORM

We all know families of long-lived individuals. But exactly how much do genes influence longevity? A study of 2,700 Danish twins suggested that the heritability of human longevity—a quantitative measure of how much differences in genes account for differences in their ages at death—was only about 25 percent. Further, these genetic factors were thought to be due to the sum of small effects from a large number of genes, and therefore difficult to pinpoint on the level of an individual gene. By the time that the Danish study was carried out in 1996, a lowly worm was already helping to overturn that idea.

That lowly worm was the soil nematode *Caenorhabditis elegans*, introduced into modern biology by Sydney Brenner, a giant of the field known for his caustic wit. Born and initially educated in South Africa, he spent much of his productive life in Cambridge, England, before he established labs all over the world from California to Singapore, leading some of us to remark that the sun never set on the Brenner Empire. He first became famous for having discovered mRNA. More generally, he worked closely with Francis Crick on the nature of the genetic code and how it was read to make proteins. Once he and Crick decided that they'd solved that fundamental problem, Brenner turned his attention to investigating how a complex animal develops from a single cell, and how the brain and its nervous system work.

Brenner identified *C. elegans* as an ideal organism to study because it could be grown easily, had a relatively short generation time, and was transparent, so you could see the cells that made up the worm. He trained a number of scientists at the MRC Laboratory of Molecular Biology in Cambridge and spawned an entire worldwide community of researchers studying *C. elegans* for everything from development to behavior. Among his colleagues was biologist John Sulston, whom you met in chapter 5. One of Sulston's more remarkable projects was to painstakingly trace the lineage of each of the roughly 900 cells in the mature worm all the way from the single original cell, which led to an unexpected discovery: certain cells are programmed to die at precise stages of development. Scientists went on to identify the genes that sent these cells to commit suicide at just the right time in order for the organism to develop.

For an animal with only 900 cells, these worms are incredibly complex. They have some of the same organs as larger animals but in simpler form: a mouth, an intestine, muscles, and a brain and nervous system. They don't have a circulatory or respiratory system. Though tiny—only about a millimeter long—nematodes can easily be seen wriggling around under a microscope. Being hermaphrodites, they produce both sperm and egg, but *C. elegans* can also reproduce asexually under some conditions. They are normally social, but scientists have found mutations that make them antisocial. Worms feed on bacteria, and just like bacteria, they are cultivated in petri dishes in the lab. They can be frozen away indefinitely in small vials in liquid nitrogen and simply thawed and revived when needed.

Worms typically live for a couple of weeks. However, when faced with starvation, they can go into a dormant state called *dauer* (related to the German word for endurance), in which they can survive for up to two months before reemerging when nutrients are plentiful again. Relative to humans' life span, this would be the equivalent of 300 years. Somehow these worms have managed to suspend the normal process of aging. There is a caveat, though: only juvenile worms can enter the dauer state. Once animals go through puberty and become adults, they no longer have this option.

David Hirsh became interested in *C. elegans* while he was a research fellow under Brenner in Cambridge, then continued working with the worms upon joining the faculty at the University of Colorado. There he took on a postdoc named Michael Klass, who wanted to focus on aging. This was at a time when aging was simply thought to be a normal and inevitable process of wear and tear, and mainstream biologists viewed aging research with some disdain. However, things were beginning to change, partly because the US government was concerned about an aging population. As Hirsh recalled, the National Institutes of Health had just established the National Institute on Aging, and at least some of his and Klass's motivation for working in the area was that they knew they stood a good chance of receiving federal funding.

Hirsh and Klass first showed that, by many criteria, worms age little if at all in the dauer state. Next, Klass wanted to see if he could isolate mutants of worms that would live longer but not necessarily go into dormancy. This would help him identify genes that affected life span. To rapidly produce mutants that he could screen for longevity, he treated the nematodes with mutagenic chemicals. He ended up with thousands of plates of worms, which he continued studying after starting his own lab in Texas. In 1983 Klass published a paper about a few long-lived mutant nematodes, but eventually he shut down his lab and joined Abbott Laboratories near Chicago. Before doing so, however, he sent a frozen batch of his mutant worms to a former colleague from Colorado, Tom Johnson, who by then was at the University of California, Irvine.

By inbreeding some of the mutant worms, Johnson found that their mean life span varied from ten to thirty-one days, from which he deduced that, at least in worms, life span involved a substantial genetic component. It still wasn't clear how many genes affected life span, but in 1988 Johnson, working with an enthusiastic undergraduate student named David Friedman, came to a striking conclusion that ran completely counter to the conventional wisdom that many genes, each making small contributions, influenced longevity. Instead, it turned out that a mutation in a single gene, which the two called *age-1*, conferred a longer life span. Johnson went on to show that worms with the *age-1* mutation had lower mortality at all ages, while their maximum life span was more than double that of normal worms. Maximum life span, defined as the life span of the top 10 percent of the population, is considered a better measure of aging effects because mean life span can be affected by all sorts of other factors that don't necessarily have to do with aging, such as environmental hazards and resistance to diseases.

At the time, Tom Johnson was not a famous scientist, and his premise that a single gene could affect aging to such a degree defied the consensus view. Thus it took almost two years for his paper to be published. Even after it finally appeared in the prestigious journal *Science* in 1990, Johnson's work was viewed with some skepticism by the scientific community.

But then, a few years later, came a second mutant worm. This effort was led by Cynthia Kenyon, already a rising star in the *C. elegans* field. Kenyon had a golden career: PhD from MIT; postdoctoral work with Sydney Brenner at the MRC Laboratory of Molecular Biology in Cambridge, where the first studies on the genetics of the worm were being carried out; faculty member at the University of California, San Francisco, another world-renowned center for molecular biology and medicine. Kenyon had established herself as a leader in the worm's pattern development, which is the process by which it lays down its body plan as it grows. She was interested in aging research, but since it was still an unfashionable discipline, she found it difficult to enlist students to work on the problem. After hearing Tom Johnson speak about his work on *age-1* at a meeting in Lake Arrowhead just outside Los Angeles, though, she felt inspired to work on the problem of aging and began her own screening for new mutants.

Like Hirsh, Klass, and Johnson, Kenyon focused on dauer formation. In the previous decade, scientists had identified many genes that affected dauer formation, usually prefixed by the letters *daf*. Scientists traditionally italicize the names of genes; when not italicized, the letters refer to the proteins that the genes encode. Under normal conditions, these mutations would predispose worms to enter the dauer state. But Kenyon had a hunch that some of these genes would affect longevity even outside the dauer state. She employed a trick in which she used mutant worms that were temperature sensitive: they would not enter the dormant state at a lower temperature (68°F, or 20°C). They were allowed to develop at this lower temperature until they were no longer juveniles and dauer formation was no longer an option. At that point, they were shifted to a higher temperature of 77°F (25°C) and allowed to mature into adulthood so that their life span could be measured.

From these studies, Kenyon and her colleagues identified a mutation in a gene, *daf-2*, that lived twice as long as the average worm. In marked contrast to the skepticism Johnson faced, Kenyon had no trouble publishing her work: her 1993 paper in *Nature* was received with great fanfare. Apart from her stellar academic pedigree and scientific abilities, Kenyon was also lucid and charismatic, so she was extolled by the media. In an unfortunate omission, neither Kenyon's paper nor the accompanying commentary mentioned Johnson's earlier work on *age-1*, and much of the reporting of Kenyon's work gave the impression that it was the first time that a mutation that extends longevity had been discovered.

At this point, nobody had any real idea of what the genes identified by Johnson and Kenyon actually *did*. Enter Gary Ruvkun. Today Ruvkun is most famous for discovering how small RNA molecules called microRNAs regulate gene expression, but he has led a varied and colorful life, both personally and scientifically. When I met him about ten years ago at a meeting in Crete, he became increasingly gregarious after a few drinks; at one point, he donned a bandanna and pretended to smoke a cigarette while pouring himself some strong Greek liquor, which, with his luxuriant but well-tended mustache, made him look like a sailor on shore leave in a Greek taverna. All the while, he incongruously continued to hold forth on RNA biology. In the mid-1990s he too was using the worm and had been studying dauer mutants, including *daf-2*, for reasons unconnected with aging. Apparently he did not hold the field in high regard, because he recollected that when Kenyon's report came out, "I thought, 'Oh, gosh, now I'm in aging research.' Your IQ halves every year you're in it."

The big breakthrough came when Ruvkun isolated and sequenced the *daf-2* gene. It coded for a receptor that sticks out of the cell's surface and responds to a molecule very similar to insulin: IGF-1 (insulin-like growth factor). Both insulin and IGF-1 are hormones that bind to their receptors in the cell. Both receptors are also kinases that activate downstream molecules, which in turn affect metabolic pathways that play a role in longevity. These hormones or their counterparts exist in nearly all organisms, so they must have originated very early in the evolution of life. That these ancient hormones control aging was a stunning finding.

These discoveries led to a general understanding of how this pathway would work. IGF-1 binds to the daf-2 receptor, which is a kinase, and activates it. This sets off a cascade of events in which one kinase acts upon another until a protein called daf-16 is phos-phorylated. It's basically the domino effect. The last domino in the chain, daf-16, is a transcription factor, so its role is to turn on genes. When it is phosphorylated, it cannot be transported to the nucleus, where the genes reside on the chromosomes, so it cannot act on its target genes. But if we disrupt the pathway—for example, by muta-tions in any of the proteins in this cascade—daf-16 can move into the nucleus and turn on a large number of genes that help the worm survive in the dauer state during stress or starvation, thus extending its life span. As it turns out, the *age-1* gene originally identified by Tom Johnson is somewhere in the middle of the cascade that starts with daf-2 and ends in daf-16.

Daf-16 turns on genes that are involved in coping with stress triggered by starvation or increased temperature, as well as genes that code for the chaperones that help proteins fold or rescue un-folded or misfolded proteins before they become a problem for the cell. Kenyon wrote in a 2010 review that these genes "constitute a treasure trove of discovery for the future." The pathway explained a puzzling paradox. Aging or longevity was thought to be the effect of a large number of genes, each of which would have a small effect. How could a mutation in a single gene, such as *age-1* or *daf-2*, effectively double the life span of the worm? Clearly the reason was that they were part of a cascade that ended up activating daf-16, which then turned on multiple genes that collectively exerted a cumulative effect on life span.

The idea that a growth hormone pathway might be involved in longevity also explains a curious fact. Larger species generally live longer than smaller ones because they have slower metabolisms and can also escape predation. But within species, smaller breeds generally live longer than larger ones. For example, small dogs can live twice as long as large dogs. This may have to do partly with how much growth hormone they make.

Remember that queen ants live many times longer than worker ants. Among the many reasons for this is that queens produce a protein that binds insulin-like molecules and shuts down the IGF-like pathways in ants.

But what of quality of life? Are these long-lived worms sickly and barely surviving? In a word, no. The nematodes don't just live longer, they look and act like much younger worms. We all know that one of the horrors of aging is the onset of Alzheimer's disease. Researchers can generate a model for Alzheimer's disease by making a genetic strain of worms that manufactures amyloid-beta protein in their muscle cells, paralyzing them. However, if the experiment is repeated—but this time using a strain of long-lived worms with mutations in the IGF-1 pathway—paralysis is reduced or delayed. Thus, the same mutations that extend life may also protect you from Alzheimer's and other age-related diseases that are caused by proteins misfolding and forming tangles. In fact, these mutations may prolong life precisely *because* they protect against some of the scourges of old age.

It is all very well to make worms live longer and healthier, but what about other species? Evidence elsewhere in the animal kingdom suggests similarly a strong relationship between the IGF-1 pathway and life span. Deleting the gene that codes for a protein called CHICO, which activates the IGF-1 pathway in flies, made them live 40–50 percent longer. They were significantly smaller but seemed healthy otherwise. The IGF-1 receptor is essential, but mice, like humans, have two copies of it (from their maternal and paternal chromosomes), and knocking out one of them made the mice live longer without any noticeable ill effects.

Scientists, of course, are not doing all this work to help mice. We want to know what happens in humans, but you can't just mutagenize people. There are people who naturally have mutations in the insulin receptor. Some of them suffer from a disease called leprechaunism, which stunts growth, and seldom reach adulthood. An analysis of subjects with the disease showed that the same mutations in *daf-2* would affect dauer formation in the worm, yet the consequences were rather different. Still, there are hints that this pathway plays a role in human longevity. Mutations known to impair IGF-1 function are overrepresented in a study of Ashkenazi Jewish centenarians, and variants in the insulin receptor gene are linked to longevity in a Japanese group. Variants in proteins identified as part of the IGF-1 cascade have also been associated with longevity. It may be tempting to see the IGF-1 and insulin pathway as a straightforward route to tackling aging. But just the complexity of the pathway and the range of effects it produces tells us it is a finely tuned system, and tinkering with it while avoiding unforeseen ill effects could be difficult.

When food intake is restricted, the levels of both IGF-1 and insulin decline. If the IGF-1 pathway is inhibited already, you might not expect caloric restriction to have much additional effect. Exactly as you might predict, caloric restriction did not further increase the life span of daf-2 mutant worms; moreover, its full effect depended on daf-16. But this too is puzzling, because the other, completely different TOR pathway is also affected by caloric restriction. So even if the IGF-1 pathway was disrupted, shouldn't caloric restriction have had at least some effect through the TOR pathway? It turns out that these two pathways are not completely independent. They are two large hubs in a large network, but there is lots of cross talk between them. In other words, proteins that are activated as part of one pathway will activate ones in the other pathway, so they are interconnected. In particular, TOR is activated by elements of the IGF-1 pathway as well as by nutrient sensing.

While the two pathways are highly coordinated, they are not the whole story behind caloric restriction. Two scientists found a mutant that causes partial starvation of the worm by disrupting its feeding organ, the equivalent of the throat. The mutant, *eat-1*, lengthens life span by up to 50 percent and does not require the activity of daf-16. Also, double mutants of daf-2 and eat-1 live even longer than the daf-2 mutants alone. This means that caloric restriction affects other pathways besides TOR and IGF-1.

Mutations that affect longevity dramatically might seem to suggest that aging is under the control of a genetic program. This idea might seem to contradict evolutionary theories of aging, but, in fact, it doesn't. When worms were subjected to alternative cycles of food and scarcity, it turned out that the long-lived mutant worms simply could not compete reproductively with shorter-lived, wild-type worms. These pathways allow organisms to have more offspring at the cost of shortening life later on, exactly as one might predict from the antagonistic pleiotropy or disposable soma theories of the evolution of aging.

We have seen what rapamycin can do, but is there a drug that acts elsewhere, such as on the IGF-1 pathway? There is a great deal of interest in metformin, a diabetes treatment. Diabetes, of course, is related to deficient insulin secretion or regulation rather than to IGF-1, although the two molecules are closely related. To understand the difference between these two hormones, I took a short walk from my own lab to the nearby Wellcome-MRC Institute of Metabolic Science on the Addenbrooke's Biomedical Campus in Cambridge, England, to meet Steve O'Rahilly, one of the world's experts on insulin metabolism and its consequences for diabetes and obesity.

Despite his many distinctions and his job as the director of a major institute, Steve lacks even a hint of self-importance. He is a jolly man who in his talks often jokes that his physique makes him particularly qualified to study obesity and its causes; while far from obese, he certainly looks well fed. But underneath the jovial demeanor, he is a sharp and critical scientist who has advanced a messy field by imbuing it with intellectual rigor. Among his many contributions is demonstrating the importance of appetite genes in obesity. Here too Steve has a highly personal interest: he told me that appetite can be such a strong urge that when he is hungry, he can hardly concentrate on anything besides food.

Steve pointed out that while insulin and IGF-1 are similar in structure and have similar effects when they act on the cell, they have some major differences. Insulin has to act very quickly and in just the right amounts. Getting insulin regulation wrong can be lethal. The brain needs glucose for fuel, so hypoglycemia, a drop in blood sugar caused by too much insulin in the circulation, is very dangerous even if it only lasts a few minutes.

Insulin receptors are particularly abundant in liver, muscle, and fat cells. In the fasting state, insulin levels are relatively low, and the liver produces the glucose needed constantly by the brain from stored carbohydrates and other sources. But even that low level of insulin is needed to prevent the liver from making too much glucose or ketone bodies (a product of metabolizing fat). After a meal, the level of insulin surges by between ten- and fifty-fold, promoting the uptake of glucose into muscle cells, the synthesis of lipids (fat) in the liver, and the storage of lipid in fat cells.

Newly secreted insulin does not last long in the bloodstream, with a half-life of only about four minutes. If insulin is like a speedboat racing to its destination, IGF-1 is more like an oil tanker. Its effect lasts much longer, and, in the circulation, it is often bound to other proteins and not active. It needs to be released from them to act, and exactly how this happens is not clear, but that too may be under hormonal control. Also, unlike insulin receptors, IGF-1 receptors are distributed much more broadly throughout all the cells in the body, and there are more of them during development, when the organism has to grow.

IGF-1 is produced in response to the secretion of growth hormone, but its action controls the amount of growth hormone in a complicated feedback loop. When IGF-1 levels are low or IGF-1 is defective, the body responds by producing more growth hormone. The problem is that growth hormone has other effects apart from stimulating the production of IGF-1. Most notably, it releases fat from fat cells. Not storing away fat in these cells is the cause of much human pathology, such as clogged arteries, or messing up the metabolism in our liver and muscle. So it is not surprising that mutations in the receptor for insulin or IGF-1 can cause diabetes. On the other hand, with caloric restriction, you are consuming the bare minimum of calories. So you actually have less spare fat because you are burning it off to provide energy. This means that caloric restriction does not have the same consequences as simply reducing the level of IGF-1, where excess fat is released to cause damage. Because of this fundamental difference, drugs that try to mimic caloric restriction by acting on the IGF-1 pathway could be particularly challenging to develop. It is hard to cheat our bodies' finely tuned system.

That is what explains the current interest in metformin. The drug is already used by millions of people with diabetes all over the world, so it has gone through various clinical trials for safety. Its use, in fact, dates all the way back to medieval Europe, where extracts of the plant *Galega officinalis*, commonly known as French lilac or goat's rue, were used to relieve the symptoms of diabetes. One of the products of the extract, galegine, could lower blood glucose but was too toxic. Eventually a derivative, metformin, was synthesized and tested and is now the first-line treatment for type 2 diabetes, which is more common later in life and is caused not by a lack of insulin but because the insulin doesn't bind well to its receptor.

How metformin works as a treatment for type 2 diabetes is not entirely clear. Traditionally, most charts of metformin interactions resemble an incredibly complicated wiring diagram. Because of recent advances in our ability to visualize biological molecules, we can now see exactly how metformin binds and inhibits its target protein.

This target protein is a crucial component in the process of respiration, in which oxygen is used to burn glucose to produce energy in our cells. Disrupting our ability to utilize glucose in turn affects our energy metabolism and acts on components of the IGF pathway, including an enzyme that regulates glucose uptake. Although some studies have claimed that metformin reduces glucose production in the liver, others show that it actually increases it in healthy people and those with mild diabetes. According to another study, the drug alters our gut microbiome in a way that is at least partly responsible for its effects. Steve O'Rahilly's work demonstrates that metformin also works by elevating the levels of a hormone that suppresses appetite.

It may seem odd that a drug whose mode of action is so complex and poorly understood should be so widely prescribed for people with diabetes, but this is often the case in medicine. For almost a hundred years, we had no idea how aspirin worked, yet people consumed billions of tablets for their aches and pains. Still, given the uncertainties, it is rather surprising that metformin has now become interesting as a potential drug to combat aging. This is partly because of a couple of early studies. In the first, from the National Institute on Aging, long-term treatment with metformin in mice improved both their health and life span. A second study, in humans, showed that diabetics on metformin lived longer not only than diabetics on other drugs but also longer than nondiabetics—a significant finding, since diabetes itself is a risk factor for aging and death.

Such promising outcomes certainly raised optimism about using metformin to prolong healthy life even in people without diabetes, but subsequent studies have questioned these results. One, from 2016, concluded that metformin was merely better than other diabetes drugs, so that diabetics on metformin had about the same survival rate as the general population. More than metformin, it was the family of cholesterol-lowering medications known as statins that dramatically reduced mortality, especially in patients with a history of cardiovascular disease. Metformin did extend the life of worms if treatment was initiated at a young age, but it was highly toxic and actually shortened life span when treatment commenced at an older age. Curiously, some of the toxicity was alleviated by giving the worms rapamycin at the same time. Metformin also undermined the health benefits of exercise, which itself is well established as one of the best remedies against diseases of aging. And one study claimed that diabetics on metformin exhibited an increased risk of dementia, including Alzheimer's disease.

Given these uncertainties, Nir Barzilai, a gerontologist at Einstein College of Medicine in New York, is the principal investigator for a large clinical trial of about three thousand volunteers between the ages of sixty-five and seventy-nine called Targeting Aging with Metformin (TAME). The study's goal is to see if metformin delays the onset of age-related chronic diseases such as heart disease, cancer, and dementia, as well as monitor for adverse side effects.

To date, however, despite considerable effort, the evidence for metformin concerning longevity is not at all clear. Its effect isn't nearly as strong or as well established as that of rapamycin, which inhibits the TOR pathway. One reason for the interest in metformin is that its long-term safety has been established in diabetics. Those with diabetes will be perfectly happy to take metformin, as their risk of poor health and eventually dying of complications of diabetes is much higher without treatment. But given the potential drawbacks noted here, it is quite a different matter to recommend its long-term use in healthy adults just yet.

WE HAVE COME A LONG way from the age-old idea that exerting self-control over one's diet is good for you and that gluttony comes at a steep price to our health. First there was the scientific evidence that caloric restriction could prolong healthy life compared to an ad libitum diet. Then in the last few decades, two previously unknown pathways, the TOR and the IGF-1, were shown to be major processes in the cell that responded to caloric restriction. This in turn has opened up the possibility of extending healthy living and even life span by tinkering with these pathways. The world of medical science has compiled a tremendous amount of research regarding the effects of rapamycin, metformin, and related compounds on aging and life span; rapamycin and its chemical analogs are among the more promising avenues for tackling aging. Still, bear in mind that inhibiting these pathways individually is not the same as caloric restriction, and a lot more work needs to be done to establish both the efficacy and safety of these approaches.

Several things strike me about the discovery of TOR and the IGF-1 pathways. First, the mere existence of these pathways came as a complete surprise. Second, at least in the case of TOR, scientists were not even looking originally for a connection with caloric restriction, let alone aging. By sheer chance, they uncovered major processes in the cell that have ramifications not only for aging but also for many diseases. Third, they involved organisms that might not seem obvious for studying aging, such as yeast and worms. Finally, the discovery that a single gene could impact life span so dramatically was quite unexpected.

Before we leave the complicated maze of caloric restriction and its pathways, let us visit a third strand that, like the story of TOR, begins with baker's yeast. Unlike the discoverers of TOR, who were not even investigating anything pertaining to the aging process, this story is about scientists who deliberately used yeast to discover genes related to aging. A yeast cell divides by budding off smaller daughter cells. The mother cell acquires scars on its surface with each budding and can only undergo a finite number of divisions. This inability to divide further is called replicative aging. Still, you might not think that studying this rather specialized property of a single-celled organism such as yeast would have any relevance at all for a phenomenon as complex as human aging. That was exactly the skepticism that Leonard Guarente encountered from his colleagues at MIT when he said he was planning to tackle aging using yeast.

Like many molecular biologists, Guarente had relied on yeast to study how genes are turned on and off by controlling the transcription of DNA into mRNA. By 1991, three years after Johnson's report on the long-lived *age-1* mutant in worms, Guarente was a tenured faculty member at MIT. He was already established and professionally secure, so when two of his students, Brian Kennedy and Nicanor Austriaco, told him they wanted to work on aging, Guarente agreed to embark on what for him was an entirely new area, dramatically altering the trajectory of his career.

Initially, Guarente and his students identified a trio of genes belonging to a family called SIR genes, for silent information regulator. The SIR family in turn controls genes that define the mating type or "sex" of yeast. (Yeast mating is complicated, and they can switch their "sex" from one type to another.) Eventually Guarente's team showed that just one of these genes, *Sir2*, had the biggest effect on yeast life span. Increasing the amount of Sir2 in cells extended life span, while mutating it reduced life span. The effect was not as large as the factor of 2 seen for the *age-1* or *daf-2* mutants in worms. But they had clearly identified a gene in yeast that controlled how many times a mother cell could divide before it was exhausted. Even more promising, *Sir2* was a highly conserved gene: it had counterparts in other species, including flies, worms, and humans. They soon found, with mounting excitement, that increasing the amount of Sir2 in flies and worms also extended their lives.

But how did it work? Recall that our genome can be recoded using epigenetic marks—chemical tags—on either the DNA itself or on the histone proteins tightly associated with it. In general, adding acetyl groups to histones activates those regions of chromatin, whereas removing acetyl groups silences them. Sir2 turns out to be a deacetylase, which you might recall are enzymes that remove acetyl groups from proteins such as histones, and there is evidence that this activity silences genes near the boundary of telomeres and affects life span. Sir2 also requires a molecule called nicotinamide adenine dinucleotide (NAD), which is required for metabolizing energy in the cell. This was a hint that when there is starvation, there is not enough free NAD to activate Sir2. Suddenly you could make a plausible link between Sir2 and caloric restriction, which had long been implicated in aging in many organisms, including yeast. Sure enough, in both flies and yeast, mutation of Sir2 eliminated the benefits of caloric restriction in prolonging life, and, in worms, the effect of Sir2 required the presence of daf-16, the same transcription factor that had already been identified as the target of the IGF-1 pathway in worms. Suddenly things appeared to come together: a mutant affecting life span in yeast was associated with a pathway affecting aging in worms that in turn was connected with caloric restriction.

Finding mutants that increased longevity in both worms and yeast prompted Guarente and Kenyon to publish a highly enthusiastic article in the journal *Nature* extolling the prospects of curing the aging problem. "When single genes are changed," they wrote, "animals that should be old stay young. In humans, these mutants would be analogous to a ninety-year-old who looks and feels forty-five. On this basis, we begin to think of ageing as a disease that can be cured, or at least postponed." They went on to found a company in Cambridge, Massachusetts, with the equally optimistic name Elixir Pharmaceuticals.

Not long after Guarente had made his initial breakthrough, he gave a talk in Sydney, Australia. In the audience sat David Sinclair, a brash young graduate student working on his PhD at the University of New South Wales. Sinclair was clearly both impressed and excited by Guarente's results because he persuaded the latter to take him on as a postdoctoral fellow at MIT. Following his fellowship, Sinclair started his own lab at Harvard Medical School, across the river in Boston, and continued to work on Sir2 and aging, in effect becoming a competitor of his former mentor. Next, Sinclair started his own company, bearing the more descriptive and modest name of Sirtris Pharmaceuticals.

By then, researchers were keen to see if the counterpart of Sir2 in humans and other mammals would have similarly beneficial effects on life span and health. In mammals, there are seven members of this family, numbered SIRT1 through SIRT7. These proteins, like the equivalents of Sir2 in other organisms, were collectively called sirtuins. (Proteins that activate other proteins are often given names ending in *in*; *sirtuins* is simply a play on "Sir2-ins." SIRT1 seemed the most similar to Sir2, so it drew the bulk of early attention. The goal was to find a pill—or magic elixir—that would activate sirtuins in some beneficial way.

Here the story takes a rather strange, and rather French, turn. It has long been speculated that the French have a relatively low prevalence of heart disease despite their rich diet because they also drink copious quantities of red wine. Sinclair, collaborating with a biotech company in Boston, identified resveratrol as one of the compounds that stimulated SIRT1. Oenophiles around the world rejoiced, for resveratrol was a compound present in red wine. Finally, here was scientific evidence for the benefits of a French lifestyle. Their enthusiasm was apparently not tempered by the realization that it would take about a thousand bottles of wine to produce the amount of resveratrol used as a dose in those studies.

Sinclair's team and a competing group appeared to clinch the issue when they administered resveratrol to mice fed a diet high in sugar and fat. Although the mice remained overweight, and their *maximum* life span was unaffected, they were protected against the diseases of overeating: more of them survived to old age, and their organs were not diseased like those in typically obese mice.

This seemed exactly the Get Out of Kale Free card people were waiting for: permission to overindulge on an unhealthy diet without any ill effects. Never shy when it came to self-promotion, Sinclair was all over the news again when the pharmaceutical giant GlaxoSmithKline bought Sirtris for an astonishing $720 million in 2008. He had hit both the scientific and commercial jackpots—or so it seemed. But even at the time, there was considerable skepticism in the industry about the purchase.

There has been significant pushback against the claims made by sirtuin advocates, some of it coming, oddly enough, from two of Sinclair's former colleagues in the Guarente lab: Brian Kennedy and Matt Kaeberlein. Among other things, their work showed that contrary to earlier findings, caloric restriction results in an even greater life span extension in yeast cells lacking Sir2, suggesting that the two were not likely to be linked. Rather, Sir2 may have been acting in other ways by modifying the program of gene expression by deacetylating histones on DNA. The two went on to reveal that the activity of resveratrol on SIRT1 was due to the presence of a fluorescent molecule that was used to detect the activation. Without this additional molecule, no increase in activity was observed, so it was not even clear whether resveratrol had any effect on SIRT1. Not only that, but they did not find any effect of resveratrol on Sir2 activity in yeast, including life span. Pharmaceutical companies do not usually spend time proving one another wrong, but in an unusual step, scientists at Pfizer published a report stating that several of the other compounds identified by Sirtris did not directly activate SIRT1 either.

With any machinery, it is much easier to do something that will stop it from working than to improve its performance. It is the same with drug development; many drugs work by inhibiting an enzyme, and manufacturing a new drug that makes an enzyme more effective is always a challenge and relatively rare. So Glaxo's very expensive purchase of Sirtris raised eyebrows in the industry. Eventually it gave up on the lead compounds it had acquired from Sirtris and shut down the division. Five years after the sale, an article in *Forbes* magazine concluded that the best way to experience the benefits of red wine was to drink it in moderation.

Of course, following the dictum of the German theoretical physicist Max Planck that scientists rarely change their minds in light of contradictory evidence, Sinclair and others stuck to their guns. They countered the new findings by reporting that resveratrol worked alongside other helper compounds in the cell that had properties similar to the fluorescent molecules they had used to monitor Sir2 activity in the test tube. This led to another commentary, this time in the journal *Science*, titled, "Red Wine, Toast of the Town (Again)."

However, this optimistic assessment must be weighed against a systematic 2013 study by the National Institute on Aging that evaluated several compounds proposed to increase healthy life or overall life span, including resveratrol. None of them had any significant effect on the longevity of mice. Among the others were curcumin, which is present in the herb turmeric, and green tea extract—not that these findings seem to have put many health food stores out of business.

Beyond resveratrol, skeptics began to question the very premise of the sirtuin idea. Sir2 extends replicative life span, but losing the ability to keep reproducing is only one kind of aging in yeast. There is also chronological life span, which measures how long yeast can survive in a semi-dormant state—for example, when it has run out of nutrients. Sir2 activation actually reduces chronological life span in yeast. We humans—with the exception, perhaps, of a few very rich old men—are not mainly concerned with our ability to reproduce in old age, but with increasing life span and improving health.

Later studies also contradicted some of the early studies about the effect of Sir2 on life span. If you ascribe an effect to a mutation, you need to take care that in creating the mutant strain, you have not changed any of the thousands of other genes in the organism. Scientists clarified that overproduction of Sir2 in worms and flies had no effect on the life span of either worms or flies as long as they did not change anything else about the genetic makeup of their organisms. This considerably deflated enthusiasm for sirtuins as a potential boon to extending life, as illustrated by journal articles titled "Midlife Crisis for Sirtuins" and "Ageing: Longevity Hits a Roadblock." Feeling embattled, Leonard Guarente repeated the experiment in worms by overproducing Sir2 without changing the genetic background, and had to revise his previous estimate of an up to 50 percent increase in life span down to about 15 percent.

The sirtuin with the most dramatic effect may actually turn out to be SIRT6; mice deficient in SIRT6 develop severe abnormalities within two to three weeks and die in about four weeks. The protein is also a histone deacetylase that may affect how genes are expressed in telomeric chromatin, and some studies suggest that it increases life span in mice, with one study theorizing it does so because it stimulates DNA repair.

It is telling that two of the pioneers of sirtuins in Guarente's own lab, Kennedy and Kaeberlein, both well-established, respected researchers in their own right, have now entirely moved away from sirtuins to focus on other aspects of aging research such as the TOR pathway and how rapamycin affects it. Sirtuins, through their action on histones, may be involved in patterns of gene expression and genome stability, and are important for human physiology in ways that still need to be understood. But enthusiasm for their use in aging has declined except among the faithful. Many in the gerontology community are highly dubious that they have any direct connection with caloric restriction or extension of life span.

There is one related molecule that has retained considerable prominence regardless of the fate of sirtuins: NAD. Nicotinamide adenine dinucleotide plays many essential roles in the cell, including for sirtuin function. It is made by the body using nicotinic acid (niacin) or nicotinamide, both slightly different forms of vitamin B_3, although it can also be made by our cells from the amino acid tryptophan or by salvaging some recycled molecules.

In the cell, NAD cycles between an oxidized and reduced form to help our cells burn glucose to convert it into other forms of energy. This process, called respiration, is absolutely essential for our ability to use glucose as a fuel; however, it does not use up NAD rapidly, since it simply cycles back and forth between its two forms. But NAD performs other essential functions, such as repairing DNA and altering gene expression through sirtuins, and these functions deplete it. Thus, as we grow older, our levels of NAD decline. The brain is one of the body's biggest consumers of glucose as a source of energy, and you can imagine how a decline in NAD levels might harm brain function. It can also cause a host of other problems, from increased inflammation to neurodegeneration. If that seems a lot for a single molecule, it simply says something about how central NAD is to our metabolism.

Our cells can't take up NAD directly from our diet. But we can utilize molecules that are direct precursors of NAD, of which two popular ones are called NR (nicotinamide riboside) and NMN (nicotine mononucleotide). Search for them on the internet, and you will find countless websites arguing that one or the other is better as an anti-aging supplement depending on which one they are selling. According to one study, increasing NAD levels by providing NR or NMN to mice slowed their loss of stem cells and protected them from muscle degeneration and other symptoms of decline; in another report, higher NAD levels led to an increase in life span. However, since NAD is so central to the chemistry of life, it may have benefits that have nothing to do with an increase in life span. Indeed, Charles Brenner, a longtime expert on NAD metabolism, says, "I expressly tell people NR is not a life extension drug and that the case for its use has nothing to do with sirtuins and everything to do with acute or chronic losses of redox [reduction/oxidation reactions involved in respiration] and repair functions in the conditions that attack the NAD system. The NR trial I am most interested in is promoting healing from scratches and burns." The results of taking either NR or NMN in humans are not yet definitive, and so far there have been no long-term studies in humans on their benefits or side effects. However, this has not stopped them from being heavily marketed as anti-aging nutraceuticals, or dietary supplements with real or alleged physiological benefits that don't require approval from agencies like the FDA. Global sales of NMN register about $280 million annually and are forecast to reach almost $1 billion by 2028.

We have seen how our cells orchestrate a finely tuned protein production program—and how this program starts to wobble as we age. A simple corrective—restricting our calories and eating well—can do much to slow this deterioration through complex interconnected pathways. Much excitement in aging research is about the prospect of producing drugs that inhibit these pathways and produce the benefits of caloric restriction.

The cell, though, is not merely a bag of proteins. It contains large structures and entire organelles that must work together in harmony. When and why those relationships break down is a topic at the forefront of aging research. And it all comes back, strangely enough, to an ancient parasite. We normally think of parasites as harmful, but this one was a mixed blessing. On the one hand, it enabled us to evolve from small unicellular organisms into the complex creatures we are today. On the other hand, it is also a major reason why we age.

9.

THE STOWAWAY
WITHIN US

A couple of times a year, I visit my ten-year-old grandson in New York and experience something that must be familiar to all grandparents. Although I am physically fit for my age, I am exhausted after spending a day with him. How does he have such boundless energy that just watching him makes me tired? One reason I lack his energy also explains why we both exist as complex creatures, and it dates back to an event that occurred about 2 billion years ago.

The earliest life forms were single-celled creatures swimming around in a primordial soup. How did they become us? Each cell in our body is much larger and more complex than a typical bacterium, so even how just one of these complex cells evolved was a mystery. In the early 1900s a Russian botanist named Konstantin Mereschkowski proposed that one cell swallowed up another simpler, smaller cell. On its own, this was not remarkable; normally, either the smaller cell was killed and digested, or the cell doing the swallowing bit off more than it could chew and perished from the indigestion. But in one such case, Mereschkowski proposed, the swallower and swallowed both survived—and have continued to coexist and replicate ever since.

The theory hung around for decades but really gained credence in the 1960s when a biologist named Lynn Margulis began
working on the idea. Margulis was an iconoclast. She was married
to the astronomer Carl Sagan before marrying Thomas Margulis,
a chemist, whom she also soon divorced, and is quoted as saying,
"I quit my job as a wife twice. It's not humanly possible to be
a good wife, a good mother, and a first-class scientist. No one
can do it—something has to go." One of her more controversial
theories is the Gaia hypothesis she proposed with scientist James
Lovelock, which states that the entire biosphere—the Earth,
its atmosphere, geology, and all the life forms that inhabit it—
is a self-regulating, living organism. She also had more extreme,
and troubling, views. Margulis wrote an essay suggesting that the
9/11 attacks on the World Trade Center were part of a conspiracy
orchestrated by the US government, and questioned whether the
human immunodeficiency virus (HIV) was really the cause of
acquired immunodeficiency syndrome, or AIDS. Her view of herself as a maverick may have attracted her to conspiracy theories,
but this attitude also allowed her to make a major contribution to
our understanding of life.

Margulis believed that symbiosis was widespread and that
eukaryotes—more complex cells that have a nucleus—evolved as a
result of symbiotic relationships among bacteria. At the time, the
dogma was that simpler bacteria evolved slowly into more complex
forms of cells. You could think of Margulis's idea as an extension
of the one Mereschkowski had proposed almost six decades earlier,
but it was still sufficiently controversial that her work was rejected
by fifteen academic journals before being published in 1967 by
the *Journal of Theoretical Biology* (under the byline Lynn Sagan).
Margulis proposed that the descendants of the bacteria that were
swallowed up now exist as organelles in the larger cell. In animal
cells, we know these as mitochondria. In addition to mitochondria,

plants have another bacterial descendant inside them: chloroplasts, which turn sunlight into sugar through photosynthesis. Neither we nor plants can exist without these stowaways inside us.

Today scientists believe that the key event that led to the formation of eukaryotes occurred about 2 billion years ago, when a single-cell organism called an archaeon swallowed a smaller bacterium. Against the odds, the bacterium survived, and eventually entered into a symbiotic relationship with its archaeon host. In the intervening 2 billion years, the bacterium evolved into mitochondria. In the 170 years since mitochondria were first discovered, scientists have learned that they are highly specialized centers of energy production in the cell. It is that ability to generate energy that allowed our primitive ancestor to evolve into today's huge and complex variety of cells and spurred the growth of complex life forms. But we also know that energy is conserved and cannot be created out of nothing. So what does it mean to say that mitochondria generate energy?

Contrast today's world with a primitive, preindustrial one. In a primitive world, there were many different sources of energy. You could use the energy of the sun to warm things; you could burn wood and other fuel to generate heat; you could use the flow of a river or the power of wind to turn a mill wheel; or use wind to sail across oceans. However, these different sources of energy are not interconvertible, and they can be used only in very limited ways. You could not, for example, use wind to cook your food.

Now think of today's world: virtually every source of energy, from solar and wind, to fossil fuels and nuclear fission, can be converted to electricity. Electricity in turn can be used for almost everything. It provides heat and light, moves us around in cars and trains, entertains us through our television sets and other gadgets, and enables instant communication around the world. Electricity has become the universal currency of energy, in much the same way that monetary currency replaced barter trade hundreds of years ago.

That is exactly what mitochondria do in a cell. They take less versatile forms of energy—for example, the carbohydrates that we consume—and convert them into the universal energy currency of the cell, which is the molecule adenosine triphosphate, or ATP. We have come across ATP before: it is one of the building blocks of RNA and consists of the adenine base attached to a ribose sugar and a string of three phosphates. The bonds between the phosphate groups are what chemists call high-energy bonds. It takes energy to form them, and that energy is released when they are broken. When the cell needs energy for any particular process in the cell, it can break the bond between the second and third phosphate groups and use the energy released as a result. ATP is like a tiny, highly mobile molecular battery.

When we digest food, especially carbohydrates, we are effectively burning the sugar that we obtain by breaking down carbohydrates. In fact, chemically it is the same as if we actually burned sugar in a flame, except that our cells do it in a very controlled way. In both cases, the result is the same: sugar combines with oxygen and releases carbon dioxide and water, and releases energy in the process. That is exactly what we do when we breathe in and out. The energy released during respiration is used by mitochondria to make ATP.

This process is chemically similar to the way we produce electricity using hydroelectric power. Unlike our own cells, which have a single membrane enveloping them, mitochondria, like their bacterial ancestors, have two membranes: each one a thin double layer of fatty molecules called lipids, which separate aqueous compartments from one another. Inside the inner membrane is a large complex of protein molecules that uses the energy of respiration to move hydrogen ions (H^+), or protons, across the inner membrane, creating

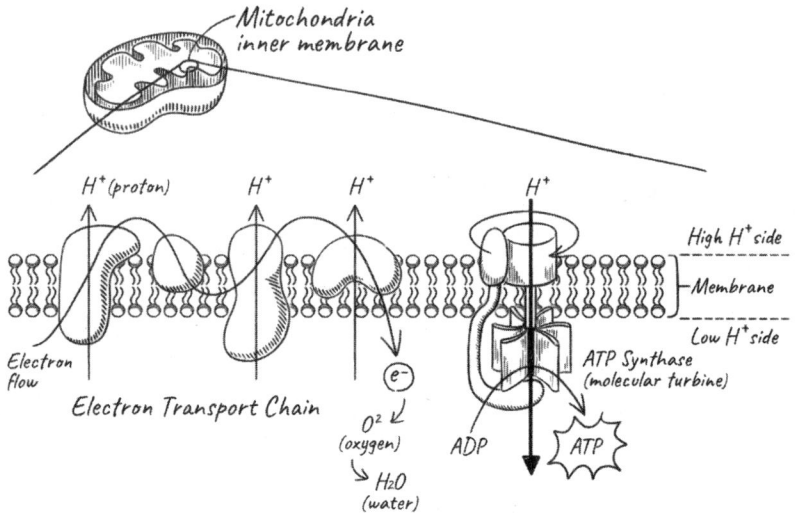

Production of energy in our mitochondria.

a proton gradient, where one side of the membrane has a higher concentration of protons than the other. And just as water flows downhill, the protons want to go down the concentration gradient. But because the membrane is not generally permeable to protons, they can do so only by traveling through a specialized molecule that acts like a molecular turbine. In the same way that water is made to go down a hydroelectric dam through large pipes to turn turbines that generate electricity, protons go through that special molecule, ATP synthase, which, as a result, actually turns like a turbine, and makes a molecule of ATP by adding on the third phosphate to adenosine *di*phosphate, or ADP, which has just two phosphates.

Just as monetary currency increased trade and prosperity dramatically, enabling complex societies to evolve, and just as the energy currency of electricity allowed societies to become incredibly complex technologically, the efficient production of ATP allowed cells to become ever more complex and specialized. ATP is a small molecule and makes its way, as needed, all over the cell. It provides the energy for everything from making the components of the cell, to moving around parts of the cell, to enabling cells themselves to move. Our muscles use ATP to generate the power to contract. In our brain, ATP maintains the voltage across membranes in our neurons while they transmit electrical signals and fire impulses. The human body has to generate roughly its own weight in ATP every day, and the brain alone uses about a fifth of that. Just thinking uses hundreds of calories a day. And mitochondria provide nearly all of that ATP.

The stowaways within us, which may well have begun their lives as parasites, have made themselves indispensable by producing the ATP we need to survive. Mitochondria differ from their bacterial ancestors in other ways too. For one thing, they've shed most of their genes, so the mitochondrial genome is now tiny, typically coding for only a dozen protein genes. More than 99 percent of the mitochondria's components are made by translating genes that now reside on the chromosomes in our nucleus. These proteins are made in the cytoplasm of our cells and then imported across one or both membranes of the mitochondria using a complicated machinery. How and why mitochondria managed to move most of their genes to their host's genome, or why they retained any genome at all, is not well understood. This small mitochondrial genome is the source of many problems, though, because mutations in the mitochondrial DNA can give rise to diseases, including diabetes, and heart and liver failure, as well as conditions such as deafness.

We inherit our mitochondria exclusively from our mothers because the sperm contributes none of its mitochondria to the fertilized egg. As a result, diseases due to defects in the mitochondrial genome are inherited entirely from the mother. A few years ago, the United Kingdom made it legal for parents to produce a "three-parent" baby. The nucleus from the egg of a potential mother with defective mitochondria is introduced into the egg of a healthy woman donor that has had its own nucleus removed. This egg is then fertilized with the father's sperm and placed in the womb of the potential mother. The child will carry mostly the genes of its father and mother, but all of his or her mitochondria, with their tiny genome, will come from the egg donor.

Cells can contain between tens to thousands of mitochondria. These mitochondria don't lead entirely separate lives as they might if they were bacteria in a culture. Rather, they are constantly fusing and splitting. Mitochondria may be fusing to intermix their contents, partly as a way to compensate for partially damaged components in each of them. They also split in different ways. When cells divide, mitochondria will also split, often down the middle. But sometimes they will also split off parts that are defective so that they can be sent off to be degraded and recycled using processes such as autophagy, which we discussed in chapter 6.

Mitochondria don't just fuse with one another; they also interact with a cell's other organelles in interesting ways. It turns out that lipids—the fatty molecules that make up our membranes—are highly specialized, so different organelles and cell types have different compositions of lipids. Mitochondria often exchange components with other organelles so that they can help one another make the specialized lipids they need. Excessive contacts between these organelles and mitochondria can be just as harmful as having too little.

Finally, they do many other things besides making ATP. For example, they are also the place where the final stages of sugar burning occurs. They are the sites of burning our stored fat, which is especially important when our carbohydrate intake is insufficient, such as when we are starving or dieting. The energy from burning fat is also used to make ATP. Beyond energy production, mitochondria are now part of a complicated signaling network with the rest of the cell. They tell the cell when energy levels are low or high, so that it can adapt accordingly by turning on or off appropriate genes and pathways.

Thus, mitochondria are no longer just energy factories but have become a central hub of the cell's metabolism, which is a far cry from the bacterial stowaway in our cells that they once were. We now coexist in a complex relationship with them. As we age, our mitochondria still work, but they have accumulated defects. Not only do they produce energy less efficiently, but they have become creakier and less effective at their myriad other tasks. Perhaps no other structure in the cell is so intimately connected to the energy of youth and the decline of the old. Aging mitochondria even acquire a different shape as they degrade, transitioning from elongated ovals to spherical blobs. You can see why my grandson, with his young, healthy mitochondria, might feel so much more energetic—and generally healthier—than I do.

IF MITOCHONDRIA ARE UNABLE TO function at some minimum level, we die. Remember, in most countries, death is defined by when our brain stops functioning. If we are unable to provide oxygen and sugar to our brain—which could be for a variety of reasons, such as a heart attack—the mitochondria in our brain tissue can no longer produce enough ATP for neurons to function, leading to brain death. A sudden loss of oxygen from a heart attack is a drastic occurrence, but even over the normal course of life, mitochondria gradually decline until they no longer function at the required level.

What brings mitochondria to this point? Mitochondria age for all the same reasons the rest of the cell does, but they have their own particular burden as well. In 1954, Denham Harman proposed something called the free-radical theory of aging. His idea was that chemically reactive species of molecules, some of them called free radicals, are produced normally as a byproduct of metabolism, and cause damage to the cell over time, accelerating aging. Harman's idea would seem to help explain the benefits of caloric restriction. If you eat less, you burn fewer calories every day, and you don't produce as many damaging chemical byproducts. Harman's theory also explained why animals with high metabolic rates tend to live shorter lives than those with slower metabolism.

Free radicals can be produced throughout the cell, but they and other reactive species are produced in abundance in mitochondria. A primary function of mitochondria is burning sugar by oxidizing it. The oxygen we breathe consists of two oxygen atoms bound tightly together to form the O_2 molecule. In mitochondria, this oxygen is reduced ultimately to two water molecules, each of which is H_2O. If the reduction of oxygen is not complete, the partially reduced molecules are highly reactive intermediates called reactive oxygen species, or ROS. These highly reactive forms of oxygen can damage other components of the cell, including proteins and DNA. Anyone who has ever had an old car knows what reactive oxygen can do to the chassis; in that case, the reaction is speeded up when there is common salt around, which is why cars in climates where roads are salted in the winter tend to corrode more quickly. So you can think of damage to mitochondria from oxidation as a case of our cells rusting from within.

Normally mitochondria have enzymes to scavenge away these reactive species before they cause harm, but the process is not perfect. A fraction of reactive molecules escape. Over time, they damage the molecules around them, including the proteins that make our cells work. The general breakdown in the function of the cell leads to aging. Apart from causing immediate damage, these reactive species can also affect future generations of mitochondria by damaging our mitochondrial DNA. That DNA codes for parts of the essential machinery for oxidizing sugar and generating ATP, and if it acquires too many mutations, the machinery produced will be defective. This in turn makes the reduction of oxygen less efficient, resulting in even more reactive species, kicking off a vicious cycle. The reactive species can also diffuse to other parts of the cell and generally cause havoc. Slowly with age, mitochondria will perform less and less effectively.

Harman's mitochondrial free-radical theory didn't gain much traction at first, but a number of observations supported it. For one thing, the production of these reactive species increases with age; by contrast, the activity of the scavenging enzymes that remove them decreases with age, compounding the harm. But it wasn't clear whether these changes were simply a result of aging or whether they themselves were further driving the aging process. Strains of mice that made more of an enzyme that scavenged hydrogen peroxide lived about five months longer than average, which is quite an increase in longevity for a mouse. As recently as 2022, scientists in Germany showed that a parasite increases the longevity of its ant hosts severalfold by secreting a cocktail that includes two antioxidant proteins as well as other compounds. You may remember that germ-line cells such as oocytes boast superior DNA repair. One way they may minimize damage is by suppressing one of the enzymes that generates reactive oxygen species.

As the free-radical theory gained credibility, antioxidants took center stage. These compounds, which combat reactive oxygen species, were touted as a panacea for everything from cancer to aging. Sales of antioxidants such as vitamin E, beta-carotene, and vitamin C soared. Cosmetic companies included vitamin E, retinoic acid, and other antioxidants in their lotions and creams to keep skin youthful. People were exhorted to eat foods rich in antioxidants, such as broccoli and kale.

Alas, although there were isolated reports of benefits from

antioxidants, an analysis of sixty-eight randomized clinical trials of antioxidant supplements, encompassing a total of 230,000 participants, suggested that not only did they not reduce mortality, but some of them—beta-carotene, vitamin A, vitamin E—actually increased it. This by itself doesn't mean that the free-radical theory has no merit. But it does mean that you cannot just pop antioxidant supplement pills and expect to get much protection against free-radical damage. Still, don't give up on the kale just yet; eating fresh fruits and vegetables is beneficial for all sorts of other reasons.

There are many potential reasons why the results from antioxidant dietary supplements have been disappointing. They may be metabolized in a way that doesn't maintain a lasting effect, or they may not properly mimic the natural process by which enzymes scavenge free radicals and reactive oxygen species. But over the last ten to fifteen years, some in the field have come to doubt that oxidative damage from reactive oxygen species and free radicals are a major cause of aging at all. Studies with other animals, including worms and flies, showed no clear correlation between the level of scavenging enzymes and life span. In fact, contrary to the report on mice I just mentioned above, studies in species as varied as yeast, worms, and mice reveal that increased levels of scavenging enzymes or other defenses don't extend life span. On the contrary, in one study, mutant worms with higher levels of free radicals lived about a third longer. Giving them a herbicide that stimulates a surge of free-radical activity prolonged their lives even more, while reducing the level of free radicals by giving the worms antioxidant supplements reduced their lives. The naked mole rat lives many times longer than other animals of the same size, yet it has higher levels of reactive oxygen species.

What could possibly be going on? This may be an example of something called hormesis, in which exposure to low levels of a toxin is actually beneficial, whereas those same toxins are harmful at higher levels. Or, as the German philosopher Nietzsche said, that which does not kill us makes us stronger. Free radicals and reactive oxygen species send signals to stimulate the production of detoxification enzymes and repair proteins, which actually have a protective effect. Moreover, these reactive oxygen species have widespread roles as signaling molecules that convey the state of mitochondria to other parts of the cell.

So if free radicals and reactive oxygen species are by them-selves not the major problem, what else about mitochondria might make them factors in aging? We know that mitochondrial DNA mutations increase with age, and accumulation of these mutations is correlated with disease. But does it cause aging? One way to settle this was to genetically engineer strains of mice in which the DNA polymerase enzyme that replicates mitochondrial DNA was made more error prone; consequently, mutations would accumulate at a much faster rate. These mutator mice were apparently normal at birth, but they soon showed many of the symptoms of premature aging, including gray hair, hearing loss, and heart disease. At the age of about sixty weeks, most of them were dead, while normal mice were still alive. This is strong evidence that damage to our mitochondrial DNA is an important factor in aging. Tellingly, these mutator mice did not have a higher level of reactive oxygen species, so it was not as if increased mutations led to defective enzymes, which then worsened the problem by accumulating reactive oxygen species. The ultimate reason these mutator mice age rapidly is still not settled. There are reports of a complicated interplay between errors in mitochondrial DNA and the stability of the bulk of the genome in the cell's nucleus, which can cause all of the more general problems associated with DNA damage.

There is no question that damage to mitochondria is bad for the cell and accelerates aging, but it is remarkably difficult to tease out the precise sources of damage. Each human cell can house tens to thousands of mitochondria, each with its own genome. So if some of them acquire serious errors in their DNA, there will still be lots of healthy mitochondria to keep the cell working. But at some point, a threshold is reached where there are simply too many defective mitochondria in the cell, which cause so many problems that they overwhelm the good mitochondria. There are also situations where some of these defective mitochondria can multiply more quickly because they don't actually do much of the work that healthy mitochondria do. In these cases, clones of these defective mitochondria can dominate, leading to serious problems for the cell.

Mitochondria are not just energy factories but also are intimately involved in the cell's metabolism. So as they acquire defects with age, they contribute to the decline of the cells they inhabit and speed up aging. The effect is most pronounced when they contribute to the decline of stem cells, because those cells play such important and diverse roles: when they become dysfunctional, they not only fail to regenerate tissue but also cause cellular senescence and chronic inflammation, all of which are hallmarks of aging.

One characteristic of aging is a chronic low level of inflammation, cleverly dubbed "inflammaging." Inflammaging owes its existence in part to our mitochondria's ancient bacterial origins. Older, defective mitochondria are more prone to rupture and can leak their DNA and other molecules into the cytoplasm of the cell. The cell mistakes these as coming from bacterial invaders, triggering inflammation. Our neurons, which are either very long lived or do not regenerate at all, are particularly prone to aging mitochondria. It may be one reason that our cognitive abilities decline. Neurons with aging mitochondria are also less able to use the recycling pathways to clear away defective proteins and organelles, all of which expend energy. As a result, we become more prone to dementia with age.

For all these reasons, maintaining healthy mitochondria is a key to good health. How the cell does this is closely related to some of the pathways involved in caloric restriction that we have come across already. It also uses autophagy to get rid of entire mitochondria that it deems defective, or even just defective parts of mitochondria that are broken off. This process, called mitophagy, targets the mitochondria for destruction and recycling. Some proteins can sense when things are going wrong and coat the surface of defective mitochondria with markers that signal the autophagy apparatus to target them for destruction. The same caloric restriction that increases levels of autophagy by the TOR pathway also increases levels of mitophagy.

If a cell disposes of defective mitochondria, it must replace them with new mitochondria; here too, caloric restriction plays a role. The inhibition of TOR by caloric restriction, or the drug rapamycin, shuts down the synthesis of many proteins but turns on the synthesis of other proteins involved in turning out mitochondria. In studies, the increased mitochondrial activity from this process was tied directly to longer life spans in fruit flies.

Besides TOR, other signals also stimulate production of new mitochondria. Sometimes, though, this effort is futile: if the cell senses a problem with mitochondrial function, it may simply end up making more defective mitochondria.

WHILE SCIENTISTS AND THE PHARMACEUTICAL industry strive to produce a pill that will combat mitochondrial dysfunction, there is a simple way to stimulate the production of new mitochondria, and it doesn't have to cost a penny: exercise. Physical activity turns on some of the same pathways that stimulate mitochondrial production in tissues ranging from our muscles to our brain. Exercise too is an example of hormesis. Too much exercise can be harmful, and even moderate exercise can temporarily increase blood pressure, oxidative stress, and inflammation, all of which are potentially problematic. Yet as long as the amount of exercise is not so excessive as to injure us, which depends on our health and many individual factors, it is highly beneficial. One way it spurs mitochondrial function is by generating the reactive oxygen species produced by incomplete oxidation when we breathe, which, as discussed earlier in this chapter, can be beneficial in the right amounts. Of course, exercise does far more than that and benefits us in many ways: reducing stress, maintaining muscle and bone mass, countering diabetes and obesity, improving sleep, and strengthening immunity. Add to this list the healthful effects of fresh mitochondria.

Eventually, despite the cell's best efforts to both recycle defective mitochondria and manufacture new ones, our mitochondria inexorably age, and in turn accelerate other aspects of our overall aging. If accumulated mutations in mitochondrial DNA are a factor in their aging, why does a baby—or my grandson—have healthy mitochondria? The same question we asked for us as individuals could be asked here too. Why is the clock reset at each generation? Recall that the resetting of the aging clock has a few reasons. The first is that germ-line cells that form the next generation have better DNA repair and age more slowly. The second is that the epigenetic marks on DNA get reset with each new generation when germ-line cells are formed. Unlike our nuclear DNA, mitochondrial DNA doesn't have the same sophisticated epigenetic mechanisms, but it is better repaired in germ-line cells. Moreover, there is a strong selection against mutations in mitochondrial DNA, so defective oocytes are not used for fertilization. There is also a strong selection against defective sperm and even defective early embryos, so any participants with deficient mitochondria should be weeded out. Nevertheless, selection is not perfect: at least some of the loss of fertility with age is due to aging mitochondria.

By now, it should be clear that all the causes of aging described so far are highly interconnected. We started off with perhaps the most fundamental molecule of all: our DNA, which contains the information necessary to make the thousands of proteins in a cell at just the right time and in the right amounts. That information needs to be protected against damage. Those thousands of proteins must work in harmony to ensure the functioning of a healthy cell, and the cell has many mechanisms to deal with problems as they arise. Beyond proteins, entire organelles such as mitochondria need to work in a symbiotic relationship with the rest of the cell. These mitochondria may have started off as an engulfed bacterium inside a larger ancestral cell, but today they have become a central hub in our metabolism. Any defects they acquire with age set off a whole sequence of events that themselves accelerate aging. All of these affect the aging of individual cells.

If individual cells in our body were to age or die, we would hardly notice it—after all, we have trillions of cells. But except in primitive life forms, cells don't exist in isolation. In our bodies, they have to communicate with one another, and work together as part of our tissues and organs. It is when a sufficient number of cells accumulate defects with age that the symptoms of aging manifest themselves: arthritis, fatigue, susceptibility to infection, decreased cognition, and more generally, bodies that simply do not work as well as they did in our youth. It is time to look at how the aging of individual cells leads to some of the morbidities of old age.

10

ACHES, PAINS, AND
VAMPIRE BLOOD

The coast-to-coast walk is one of the great long-distance treks in England. Starting in St. Bees Head on the west coast, it cuts through the most picturesque parts of the country before ending at Robin Hood's Bay on the east coast, near Whitby, Dracula's port of entry to England in the Bram Stoker novel. The entire walk runs about 200 miles. I figured when I finished it, I could get an "I Did the Coast-to-Coast Walk" T-shirt and disingenuously wear it in the States to impress people.

My opportunity came in the summer of 2013, when a group of friends and I set off. Everything was fine for the first week, but then my knee started to become more and more inflamed until I had to abandon the walk with only a few days to go. On my return, a surgeon looked at it and discovered a torn and inflamed meniscus, the result of moderate osteoarthritis. As soon as I had the knee repaired, my right shoulder started to ache—osteoarthritis striking again. I receive little sympathy from my similarly aged friends: aches and pains in our joints are simply part of life as we get older.

Joint pain is a symptom of just one kind of inflammation, and its causes are often physical, such as the wear and tear on the bones in the joint, which then pinch and inflame the soft tissue in it. But as we age, there is a much more pervasive yet less obvious inflammation that affects our health as well as our response to disease.

One cause of inflammation comes from cells that reach a senescent state because they have aged or become damaged. We've seen that when a cell senses DNA damage, it can do one of three things. If the damage is mild, it can turn on repair mechanisms. If the damage is more extensive, it can trigger signals that kill the cell; or it can send the cell into a senescent state, in which it is no longer able to divide. We saw an example of the latter when we discussed how cells stop dividing when the telomeres at the ends of their chromosomes shorten beyond a certain point. Whether a cell is killed off or whether it enters senescence, the purpose is the same: to prevent cells with a damaged genome from reproducing. Such cells run the risk of being cancerous; indeed, the entire response to DNA damage can be thought of as a mechanism to prevent cancer. As we saw earlier, nearly half of cancers have mutations in a single protein, p53, that plays a key role in the DNA damage response. These tumor suppressor genes can induce premature senescence to prevent cancer.

Just as evolutionary theories would predict, processes that prevent us from developing cancer early in life can become a problem later on. Our tissues, for instance, would stop functioning if their cells kept getting killed off without being replaced. And even though they are alive and present, senescent cells also lead to problems. The transition from a normal cell to a senescent cell is not clearly understood. It occurs because of extensive changes to the genetic program of the cell triggered by the DNA damage response. In their altered state, senescent cells no longer contribute to the normal functioning of the tissues they serve. If they are no longer functioning as they should, you might well wonder why cells go into senescence at all instead of simply being destroyed, and why they persist.

In fact, senescent cells often don't just sit there quietly doing nothing. They secrete molecules such as cytokines that cause inflammation and disrupt the surrounding tissue. This is by design. Senescent cells are often produced in response to injury or other damage, and the same secretions that set off inflammation also promote wound healing and tissue regeneration, while at the same time signaling the immune system to clear them from the tissue. But our immune system ages along with the rest of us, and its ability to clear senescent cells declines. As damage to our DNA accumulates and our telomeres shorten, we produce senescent cells in places where they don't serve any purpose and at a faster rate than our immune system can handle, leading to chronic, widespread inflammation.

In all of the causes of aging we have discussed so far, the processes are so complex and interconnected that it is always a problem to separate cause and effect. Here too, there is the nagging question of whether an increase in senescent cells and accompanying inflammation is just a consequence of aging or whether it accelerates aging further. This question was tackled in a key study led by Jan van Deursen, who was then at the Mayo Clinic in Minnesota. He and his team used a biomarker that identified senescent cells and devised a clever method to eliminate cells with that marker. Using mice that age prematurely—called progeroid mice—they showed that removing senescent cells delayed age-related pathologies in adipose (fatty) tissue, skeletal muscle, and the eye. Even late in life, removing senescent cells delayed the progression of disorders that had already been established. The study concluded by saying that removal of senescent cells could prevent or delay aging disorders and extend healthy life. A few years later, the same team demonstrated that mice whose senescent cells were killed off were healthier in many ways than those in whom these cells were allowed to build up. Their kidneys functioned better, their hearts were more resilient to stress, they were more active, and they fended off cancers for longer. They also lived about 20–30 percent longer.

According to a follow-up study, transplanting even small numbers of senescent cells into young mice was sufficient to cause persistent physical dysfunction, and even spread senescence throughout the tissues. With older mice, introducing even fewer senescent cells had the same effect. When researchers used an oral cocktail that selectively killed senescent cells, it alleviated the symptoms of both the young and old mice and reduced their mortality significantly.

These studies have led to an explosion of experiments examining senescent cells as they relate to aging. The selective targeting of these cells for destruction, called senolytics, is growing rapidly in popularity, both in academic research and industry. But destroying problematic cells like these is only one side of the coin. Most of our tissues are constantly regenerated, and if cells are destroyed either naturally or deliberately, they need to be replaced.

An old saw holds that the human body replaces itself every seven years; in other words, after seven years, you're an entirely new collection of cells. But this isn't strictly true. Our tissues don't all regenerate at the same rate. Some, such as blood and skin cells, are regenerated rapidly. Cuts, bruises, and minor burns will heal over quickly with new skin, and if you donate blood, your body replenishes it in just a few weeks. Other organs are renewed more slowly; for example, most of the cells in your liver are replaced within three years. Heart tissue is replaced even more slowly, with only 40 percent of its muscle cells replaced in a lifetime, which is why the damage caused by a heart attack is often permanent. And it was thought that the neurons in our brain are never renewed—that we are born with every neuron we will ever have. Recently, however, scientists have shown that some brain cells *are* renewed, albeit very slowly, at a rate of about 1.75 percent annually. Still, most of our neurons were present at birth, and the inability to replenish them is why diseases that destroy them—either suddenly in a stroke or more gradually as in Alzheimer's—are so horrific.

The majority of our cells, however, *are* replaced with some regularity, and the key actors responsible for regenerating tissue are those stem cells we discussed earlier. Remember that the ultimate stem cells are the pluripotent stem cells in the early embryo that can give rise to any tissue type in the body as they differentiate. But other stem cells are halfway down the path to development of the complete organism and can regenerate only specific tissues. As Leonard Hayflick discovered in the 1950s, the cells in most tissues can undergo only a certain number of divisions, but stem cells, because they are required for regenerating tissues, are not subject to this limit.

Stem cells that maintain and regenerate tissue must strike a delicate balance. They cannot all differentiate into the mature cells of the tissues, or there would be no stem cells left to carry on this task. And the stem cells that remain behind have to keep dividing into more stem cells to replenish the ones that *have* differentiated into specific tissue cells. As we age, our stem cells begin to lose this balance between producing more of themselves and regenerating tissue.

Stem cells do not divide and proliferate indiscriminately; rather, they are activated by specific signals that they receive when the body senses a need for tissue regeneration. These signals and their ability to activate stem cells decline with age, for the many reasons we have discussed before, including damage to our genome, and epigenetic marks that our DNA acquires with age. This is one reason our muscles, skin, and other tissues degenerate with age.

Apart from not being activated, stem cells themselves eventually suffer from DNA damage and telomere loss, and accumulate metabolic defects. Eventually they trigger a response such as the DNA damage response, which can lead to either cell death or senescence. With stem cells, death is more likely, partly because a stem cell that has damaged DNA might be too much of a cancer risk to keep around. The result is a gradual depletion of stem cells throughout the body, diminishing the ability to regenerate tissue. When our bones, muscles, and skin cannot regenerate, we become increasingly frail. A particularly significant decline is the population of hematopoietic stem cells, which give rise to all our blood cells, including the cells of our immune system. This leads to immune system decline or even immune dysfunction—something called immunosenescence, which is associated with an increase in disorders such as inflammation, anemia, and various cancers, as well as in increased susceptibility to infections.

Apart from a gradual loss in the number of stem cells, there is a problem with the remaining stem cells. During much of our life, we have a healthy diversity of cells that have acquired different mutations, making us a mosaic of genomes. As we age, our stem cells acquire mutations, some of which cause them to proliferate more rapidly. These rapidly multiplying stem cells are not necessarily the best for regenerating tissues, but because they have a growth advantage, they outcompete their counterparts. Consequently, old age leaves us with stem cells that have all descended from just a few clones. Not only are they less effective, but—of greater concern—the clonal mutants themselves can become sources of cancer.

If the number of stem cells declines with age, and those that remain are descendants of a few clones, some of which may be problematic, can we somehow reverse this process? In chapter 5 on epigenetics, I explained about how turning on just a few genes that code for the so-called Yamanaka factors can reprogram cells so that they can return to being pluripotent stem cells—and thus can again give rise to any tissue in the body. Might scientists learn to regenerate stem cells in the body and reverse some of the effects of aging?

When cells are reprogrammed fully with Yamanaka factors to

form induced pluripotent stem cells (iPS cells) and used to grow new tissues, they often produce tumors such as teratomas, which can be benign or malignant. One reason for this is that the Yamanaka factors are not *precisely* reversing the normal process of development. The truth is, we don't fully understand what they do or how, but the resulting induced pluripotent stem cells are not exactly the same as our own embryonic stem cells, which develop into our body—after all, teratomas are quite rare in normal development. Given the potential risks associated with the use of Yamanaka factors, one idea is to expose cells to them only transiently, so that they would not go all the way back to being pluripotent stem cells again, but just *part* of the way back developmentally so they would be transformed into the specialized stem cells for whichever tissue they came from. Even this transient and partial reversal could help rejuvenate tissue.

Many scientists had been working on this in cells in culture, but it wasn't clear what turning on these factors even transiently in an entire animal would do. A group led by Juan Carlos Izpisua Belmonte at the Salk Institute in La Jolla, California, did exactly this by turning on the Yamanaka factors in entire mice for a short burst. After six weeks, the mice appeared younger, with better skin and muscle tone. They had straighter spines, improved cardiovascular health, healed more quickly when injured, and lived 30 percent longer. These studies involved a special strain of progeroid mice that aged prematurely. Recently, though, both Belmonte's own group as well as groups led by Manuel Serrano and Wolf Reik, both in Cambridge, England, found that doing the same thing in naturally aged mice—as well as in human cells—induced similar effects. Not only did the animals (or cells) seem younger based on various criteria, but the epigenetic marks on their DNA, and the various markers in their blood and cells, were all characteristic of a more youthful state.

David Sinclair, who had spent much of his earlier career working on sirtuins, has also begun using the Yamanaka factors to reprogram cells. A newborn mouse can regenerate the optic nerve that transmits signals from the eye to the brain, but this ability disappears as the mouse develops. Sinclair and his colleagues crushed the optic nerves of adult mice, and then introduced three of the four Yamanaka factors. They omitted the fourth, c-Myc, because it is known to have cancer-causing properties. The factors prevented the injured cells from dying and prompted some of them to grow new nerve cells reaching out to the brain. In the same study, they introduced the three factors into middle-aged mice and found that their vision was as good as younger ones. Their DNA methylation epigenetic marks resembled those of younger animals. In another experiment, the team deliberately introduced breaks in the DNA of mice, which accelerated aging by inducing the DNA repair response. One of the effects was that the pattern of epigenetic marks in the genome were characteristic of an aged animal. All of these effects could be reversed by introducing the same three Yamanaka factors.

Stem cells have been the basis of a very large biotech industry for a long time because of the promise of regenerating new cells and tissues. But it was still quite astonishing that introducing Yamanaka factors into an entire animal, where they could affect virtually every tissue, could apparently reverse aging without any obvious ill effects, at least in the short term. For example, even though two of the three Yamanaka factors used in Sinclair's experiments are also linked to cancer, his mice were tumor free for nearly a year and a half after treatment. These studies generated huge excitement in the aging community because, unlike other approaches, which can slow down the inexorable progress of aging, these studies actually promise to reverse aging by restoring cells and tissues to an earlier state. Not surprisingly, Belmonte, Serrano, and Reik, all leading researchers originally in academic labs, were snapped up by Altos Labs, the private company set up to tackle aging, which had also snapped up Peter Walter, whom we encountered in chapter 6. We will have more to say about these anti-aging enterprises later.

BEFORE WE LEAVE THIS CHAPTER, let us turn to blood. Most of us don't think of blood as an organ in the same way that we consider the liver, kidney, heart, and brain. But perhaps we should. For in many ways, blood circulation is one of the most important systems in the body. It supplies essential nutrients, including oxygen and glucose, to the other organs, as well as disposes of their waste products. It enables our response to hormones, promotes healing by forming structures at the site of injuries, and fights off infections with the immune cells that circulate in our bloodstream. If we have old, defective blood—clonal or not—that is a problem.

The idea of living forever by drinking young blood has been around for a long time. I remember being terrified when I saw my first *Dracula* movie at the age of ten. But Transylvanian myths and Gothic novels aside, is it possible to replace old blood with young?

Parabiosis attempts to do just that, by surgically connecting the circulatory systems of two animals. Some of the earliest experiments date back to the nineteenth-century French biologist Paul Bert, who was interested in tissue transplantation rather than aging. He not only connected two rats but, amazingly, is reported to have attached a rat to a cat and successfully maintained this state for several months.

Sharing blood between two different animals, let alone different species, could obviously be problematic not only because of the possibility that one or both animals' immune systems will reject the transfused blood due to incompatibility (this is why blood donors have to be matched to recipients with compatible blood groups), but also psychological issues. Indeed, Clive McCay of Cornell University in Ithaca, New York, is quoted as saying, "If two rats are not adjusted to each other, one will chew the head of the other until it is destroyed." Nowadays the animals are inbred and matched genetically to avoid biochemical incompatibilities. Then they are socialized with each other for several weeks before attachment.

Early experiments on parabiosis probed questions such as the role that blood plays in metabolic disorders, including obesity. There were, however, some scientists, like McCay, who were looking at the effects on aging as early as the 1950s. His group found that when aged rats were joined to young ones for about a year, their bones became more similar in weight and density to those of their young partners. Other studies showed that the older partners in old-young pairings lived four to five months longer than normal, which for a two-year life span is a significant extension of life. But for some reason, these studies died out in the 1970s.

The field was resuscitated in the early 2000s when Irina and Michael Conboy, a husband-and-wife team in Thomas Rando's lab at California's Stanford University, again began pairing old and young mice. Within five weeks, the young blood restored muscle and liver cells in the older subjects. Their wounds healed more easily. The fresh blood even made their fur shinier. By the same criteria, the younger partner in each of the pairs tended to fare worse than usual; it, of course, was receiving older blood in the exchange.

Rando and his colleagues had left out of their 2013 published paper that they had also seen enhanced growth of the older mice's brain cells. We know that neurons, for the most part, do not re-generate. But these early results motivated one of Rando's Stanford colleagues, the neurobiologist Tony Wyss-Coray, to investigate the effects of parabiosis on the brain. He showed that old blood could impair memory in young animals, while, conversely, young blood could improve the memories of older animals. There was a threefold increase in the number of new neurons in the older mice. By contrast, the younger mice that received old blood from their conjoined partners generated far fewer nerve cells than young mice allowed to roam free did.

Against the centuries-old backdrop of the vampire myth, these reports captured people's imaginations. Rando and Wyss-Coray were deluged with phone calls from reporters and from the general public—some of them dubious, not to mention scary. There were reports of rich old men—and, yes, it usually seems to be men— procuring a ready supply of young blood to prolong their lives.

The scientists involved were more circumspect. In a 2013 journal article, the Conboys and Rando pointed out that even in highly inbred strains of mice and rats, the risk of parabiotic disease was as high as 20–30 percent. Moreover, it was not obvious whether all of the positive effects of parabiosis could be attributed to the blood; the older animal would have also benefited from the better-functioning organs of the younger partner, such as its liver and kidneys. To test this, the Conboys conducted a study in which they exchanged blood between two animals that were not joined. They found that the adverse effects of old blood were more pronounced than the beneficial effects of young blood.

Such cautionary views did not stop lots of companies from trying to capitalize on the hype, rushing ahead before any careful human trials were completed. One company, Ambrosia, offered blood plasma from donors aged sixteen to twenty-five for $8,000 a liter. Alarmed, the US Food and Drug Administration (FDA) issued a warning that these treatments were unproven and should not be assumed to be safe, and strongly discouraged consumers from pursuing this therapy outside of clinical trials with appropriate regulatory oversight. In response, Ambrosia stopped offering the treatment, but only briefly: the people involved soon began marketing it again under the aegis of a new but short-lived business named Ivy Plasma—before returning to its original name. Ambrosia's CEO, Jesse Karmazin, said, "Our patients really want the treatment. The treatment is available now. Trials are very expensive, and they take a really long time." Most serious scientists, including those who pioneered the discoveries, believe it is premature and potentially dangerous to offer these kinds of treatments to humans without proper clinical trials.

Beyond all the hype, Thomas Rando's initial findings set off an extensive search for specific protein factors in blood that could be related to aging. In theory, you could have factors in young blood that stimulate growth and improve function; by the same token, old blood might contain factors that made things worse. Wyss-Coray and his colleagues showed that it was both. As they described in a 2017 article in the journal *Nature*, proteins from umbilical cord plasma revitalized the function of the hippocampus—a part of the brain crucial for the formation of both episodic and spatial memory. As for old blood, they zeroed in on a protein that impaired hippocampus activity; blocking it relieved some of the adverse effects.

Of course, in the parabiosis experiments, young blood improved many organs, not just the brain. Amy Wagers of Harvard University, who was a member of Rando's original team at Stanford, screened the hundreds of protein factors in blood to pinpoint the ones more prevalent in old or young blood. A factor called GDF11 was abundant in young mice but not in old, and it could rejuvenate heart tissue. But it didn't just act on heart tissue. She and her colleagues showed that the factor reversed age-related deterioration of muscle tissue by reviving stem cells in old muscles and making them stronger. In a second study with her Harvard colleague Lee Rubin, they showed that it spurred the growth of blood vessels and olfactory neurons in the brain.

Stem cells can decline in number and lose function with age, and clearly some of the factors in blood work by reactivating them. But what about the old blood making the young mice worse off? A recent study by the Conboys and Judith Campisi, another leading aging researcher, showed that treating young mice with old blood quickly increased the number of senescent cells in their circulation. This means that senescence is not just a response to stress and damage from the environment, nor is it something that simply happens over time. It can also be induced rapidly. Clearing those senescent cells reversed some of the harmful effects of old blood on multiple tissues.

Blood need not even be from young animals to confer benefits. We saw in chapter 8 that exercise has a real benefit on many aspects of our metabolism, including insulin sensitivity and mitochondrial biology. It turns out that blood from adult mice that had been subjected to an exercise program can improve cognitive function and regeneration of neuronal tissue. Rando and Wyss-Coray showed that exercised blood can also rejuvenate muscle stem cells. Using a new way of measuring effect based on which mRNAs are made in different tissues, they showed that young blood and exercised blood act in different ways. Parabiosis from young animals reduced the activity of genes that caused inflammation, whereas exercise increased the activity of genes that decline with age. Although they both stimulated growth of brain tissue, each stimulated different types of cells.

Identifying aging factors in blood and understanding how they work is now a major area of research. Scientists hope that one day it might be possible to administer a cocktail of a few factors with real anti-aging effects. This hope is spurring not only basic research but also has resulted in the creation of many biotech companies, including ones founded by some of the pioneers in the field.

While science is advancing to find out precisely which combination of blood factors is most beneficial, some billionaires are unwilling to wait. They continue to be drawn to the Dracula-like allure of young blood. For instance, Bryan Johnson, the middle-aged tech mogul behind the company Braintree Payment Solutions, spends $2 million a year on his anti-aging regimen, which includes two dozen supplements, a strict vegan diet, and, as befits a techie, lots of data, including more than 33,000 images of his bowels. He went to Resurgence Wellness, a Texas outfit that describes itself as a comprehensive health and wellness clinic–slash-spa. There he was transfused with blood from his seventeen-year-old son, Talmage, and in turn donated his own blood to *his* father in a series of multi-generational blood exchanges that lent new meaning to "all in the family." Johnson stopped the transfusions from his son after seeing no benefits himself, but still felt that "young plasma exchange may be beneficial for biologically older populations or certain conditions."

IN THIS AND EARLIER CHAPTERS, we have covered the broad landscape of aging at various levels, from our genes, to the proteins they encode, and how they affect cells and their ability to function as part of an entire animal. These levels are all interconnected, so the state of our proteins and our cells influences how and which genes are expressed, which in turn affects them. By their very nature, the causes of aging encompass virtually all of biology, and as new areas of research emerge, we find new and sometimes surprising connections with aging. So why we age and die is an ongoing story, and this book has focused on processes of the greatest interest or promise.

The quest to defeat aging and death is centuries old, but it is only in the last half century that we have accumulated a detailed biological understanding of the processes that lead to them. That knowledge has brought about an explosion of efforts by both academic institutions and for-profit companies to combat aging. Now we come to these efforts, ranging from sound mainstream science to the wildest crackpot ideas.

11.

CRACKPOTS

OR PROPHETS?

Last Christmas, when my son's family was visiting from America, there was a special exhibition at the British Museum about the Rosetta Stone and how it led to the decipherment of Egyptian hieroglyphics. So we trudged off to London, and since it was a cold and wet day during the Christmas break, we found to our dismay that the museum was packed. After we battled the crowds milling about the exhibition, we were naturally curious to see the rest of the Egyptian artifacts in the museum, including its unparalleled collection of mummies. We went over to the long hall with cases enclosing one mummy after another. It was both thrilling and sobering. Thrilling that these mummies had been preserved for a few thousand years and were right there for us to see. Sobering that each of them represented a person who had been alive.

Their corpses, now in varied states of preservation, lay underneath the wrappings and caskets. It was a stark reminder yet again of the extent to which people will go to deny death. After all, Egyptians mummified their pharaohs so that they could arise corporeally at some point in the future for their journey in the afterworld. Surely now, a few millennia after the pharaohs and with more than a century of modern biology behind us, we would not do anything even remotely so superstitious. But in fact, there is a modern equivalent.

Biologists have long wanted to be able to freeze specimens so that they can store and use them later. This is not so straightforward because all living things are composed mostly of water. When this water freezes into ice and expands, it has the nasty habit of bursting open cells and tissues. This is partly why if you freeze fresh strawberries and thaw them, you wind up with goopy, unappetizing mush.

An entire field of biology, cryopreservation, studies how to freeze samples so that they are still viable when thawed later. It has developed useful techniques, such as how to store stem cells and other important samples in liquid nitrogen. It has figured out how to safely freeze semen from sperm donors and human embryos for in vitro fertilization treatment down the road. Animal embryos are routinely frozen to preserve specific strains, and biologists' favorite worms can be frozen as larvae and revived. For many types of cells and tissues, cryopreservation works. It is often done by using additives such as glycerol, which allow cooling to very low temperatures without letting the water turn into ice—effectively like adding an antifreeze to the sample. In this case, the water forms a glass-like state rather than ice, and the process should be called vitrification rather than freezing (the word *vitreous* derives from the Latin root for *glass*), but even scientists casually refer to it as freezing and the specimens as frozen.

Enter cryonics, in which entire people are frozen immediately after death with the idea of defrosting them later when a cure for whatever ailed them has been found. The idea has been around a long time, but it gained traction through the work of Robert Ettinger, a college physics and math teacher from Michigan who also wrote science fiction. Ettinger had a vision of future scientists reviving these frozen bodies and not only curing whatever had ailed them but also making them young again. In 1976 he founded the Cryonics Institute near Detroit and persuaded more than a hundred people to pay $28,000 each to have their bodies preserved in liquid nitrogen in large containers. One of the first people to be frozen was his own mother, Rhea, who died in 1977. His two wives are also stored there—it is not clear exactly how happy they were to be stored next to each other or their mother-in-law for years or decades to come. Continuing this tradition of family closeness, when Ettinger died in 2011 at age ninety-two, he joined them.

Today there are several such cryonics facilities. Another popular one, Alcor Life Extension Foundation, headquartered in Scottsdale, Arizona, charges about $200,000 for whole-body storage. How do these facilities work? Essentially, as soon as a person dies, the blood is drained and replaced with an antifreeze, and the body is then stored in liquid nitrogen. Theoretically, indefinitely.

Then there are the transhumanists who want to transcend our bodies entirely. But they don't want humanity as we know it to end before we have figured out a way to preserve our minds and consciousnesses indefinitely in some other form. In their view, intelligence and reason may be unique to human beings in the universe (or at least they see no evidence for extraterrestrial intelligence). To them, it is of cosmic importance to preserve our consciousnesses and minds and spread them throughout the universe. After all, what is the point of the universe if there is no intelligence to appreciate it?

These transhumanists are content to have only their brains frozen. This takes up less space and costs less. Moreover, it could be faster to infuse the magic antifreeze directly into the brain after death, increasing the odds of successful preservation. The brain is the seat of memories, consciousness, and reasoning, and that is their sole concern. At some point in the future, when the technology is ripe, the information in the brain will simply be downloaded to a computer or some similar entity. That entity will possess the person's consciousness and memories and will resume "life." It won't be limited by human concerns such as the needs for food, water, oxygen, and a narrow range of temperature. We will have transcended our bodies, with the possibility of traveling anywhere in the universe. Not surprisingly, transhumanists are generally ardent about space travel, viewing it as our only chance to escape destruction on Earth. One such proponent is Elon Musk, said to be the wealthiest person in the world, depending on the year, who is well known for his desire to "die on Mars, just not on impact." Presumably one of his first goals upon reaching the red planet will be to construct a cryonics facility.

The bad news is that there is not a shred of credible evidence that human cryogenics will ever work. The potential problems are myriad. By the time a technician can infuse the body, minutes or even hours may have elapsed since the moment of death—even if the "client" moved right next to a facility in preparation. During that time, each cell in the deceased person's body is undergoing dramatic biochemical changes due to the lack of oxygen and nutrients, so that the state of a cryogenically frozen body is not the state of a live human being.

No matter, say cryo advocates: we simply must preserve the physical structure of the brain. As long as it is preserved enough that we can see the connections between all the billions of brain cells, we will be able to reconstruct the person's entire brain. Mapping all the neurons in a brain is an emerging science called connectomics. Although it has made tremendous advances, researchers are still ironing out the kinks on flies and other tiny organisms. And we don't yet have the know-how to properly maintain a corpse brain while we wait for connectomics to catch up. Only recently, after many years, has it been possible to preserve a mouse brain, and that requires infusing it with the embalming fluid while the mouse's heart is still beating—a process that kills the mouse. Not one of these cryonics companies has produced any evidence that its procedures preserve the human brain in a way that would allow future scientists to obtain a complete map of its neuronal connections.

Even if we could develop such a map, it would not be nearly enough to simulate a brain. The idea of each neuron as a mere transistor in a computer circuit is hopelessly naive. Much of this book has emphasized the complexity of cells. Each cell in the brain has a constantly changing program being executed inside it, one that involves thousands of genes and proteins, and its relationship with other cells is ever shifting. Mapping the connections in the brain would be a major step forward in our understanding, but even that would be a static snapshot. It would not allow us to reconstruct the actual state of the frozen brain, let alone predict how it would "think" from that point on. It would be like trying to deduce the entire state of a country and its people, and predict its future development, from a detailed road map.

I spoke to Albert Cardona, a colleague of mine at the MRC Laboratory of Molecular Biology who is a leading expert on the connectomics of the fly brain. Albert stresses that, in addition to the practical difficulties, the brain's architecture and its very nature are shaped by its relationship to the rest of the body. Our brain evolved along with the rest of our body, and is constantly receiving and acting upon sensory inputs from the body. It is also not stable: new connections are added every day and pruned at night when we sleep. There are both daily and seasonal rhythms involving growth and death of neurons and this constant remodeling of the brain is poorly understood.

Moreover, a brain without a body would be a very different thing altogether. The brain is not driven solely by electrical impulses that travel through connections between neurons. It also responds to chemicals both within the brain and emanating from the rest of the body. Its motivation is driven very much by hormones, which originate in the organs, and includes basic needs such as hunger but also intrinsic desires. The pleasures our brains derive are mostly of the flesh. A good meal. Climbing a mountain. Exercise. Sex. Moreover, if we wait until we age and die, we would be pickling an old, decrepit brain, not the finely tuned machine of a twenty-five-year-old. What would be the point of preserving *that* brain?

Transhumanists argue that these problems can be solved with knowledge that mankind will acquire in the future. But they are basing their beliefs on the assumption that the brain is purely a computer, just different and more complex than our silicon-based machines. Of course, the brain is a computational organ, but the biological state of its neurons are as important as the connections between them in order to reconstruct its state at any given time. In any case, there is no evidence that freezing either the body or the brain and restoring it to a living state is remotely close to viable. Even if I were one of the customers who was sold on cryonics, I would worry about the longevity of these facilities, and even the societies and countries in which they exist. America, after all, is only about 250 years old.

Despite this, many people have bought into the idea of cryonics. In the United Kingdom, a fourteen-year-old girl who was dying of cancer wanted to have her body cryogenically frozen. She needed the consent of both parents, but they were separated, and her father, who himself suffered from cancer, and was not part of her life, was opposed. She took the matter to court, and the judge ruled that she was entitled to have her wishes followed—but they should be made public only after her death. This elicited an outcry from prominent UK scientists, who called for restrictions on the marketing of cryonics to vulnerable people.

In almost a mirror image of this case, the renowned baseball player Ted Williams wanted to be cremated. Upon his death in

2002 at the age of eighty-three, two of his three children insisted on having his remains frozen, igniting a bitter family feud. In the end, a compromise was reached: only the great athlete's head would be put on ice, so to speak.

According to press reports, well-known people who intend to be cryopreserved include entrepreneur Peter Thiel, one of the co-founders of PayPal; computer scientist Ray Kurzweil, best known for his prediction that in 2045 we will reach the singularity where machines will become more intelligent than all humans combined; philosopher Nick Bostrom, who is concerned that such machine superintelligence could spell an existential catastrophe for humans; and computer scientist turned gerontologist Aubrey de Grey. More about him in a moment.

Because the brain decays rapidly following death, many cryonics facilities recommend that their clients move somewhere nearby when it's known that the end is nigh. However, this may not be good enough. Remember that the only way cryopreservation has been shown to merely preserve connections in a mouse brain was by infusing embalming chemicals into its blood while it was still alive, in a procedure that kills the animal. In 2018, a San Francisco company called Nectome was reported to have plans to do exactly that to human beings: infusing a mixture of embalming chemicals into the carotid arteries in the neck—killing the customer immediately in the process. This would be carried out under general anesthesia, although what the embalming would do to the state of the brain was not clear. The company's cofounder claimed that this assisted suicide will be completely legal under California's End of Life Option Act. One might think that the prospect of certain euthanasia coupled with an uncertain outcome would be a tough sell, but the same article claimed that twenty-five people had already signed on as customers, and one of them was reported to be thirty-eight-year-old Sam Altman, cofounder of OpenAI, the artificial intelligence research lab that launched ChatGPT, who believes that minds will be digitized in his lifetime and that his own brain will one day be uploaded to the cloud. In response, Robert McIntyre, the founder of Nectome, said that those people were early supporters of his research and had not been promised or even offered anything, certainly not silicon-based mental immortality.

LET US MOVE FURTHER UP the plausibility scale, from cryonics to Aubrey de Grey. With his two-foot-long beard and a matching messianic zeal, de Grey looks the very stereotype of an upper-class English eccentric and has amassed a large cultlike following. He began his career as a computer scientist and, although not a professional mathematician, contributed a major advance toward solving a sixty-year-old mathematics problem. At some point, he met the American fly geneticist Adelaide Carpenter at a party in Cambridge and eventually married her. This sparked his interest in biology—in particular, the mitochondrial free-radical theory of aging. De Grey came to believe that aging was a solvable problem. He asserts that the first humans who will live to be 1,000 years old have already been born. De Grey's central idea is that if we can improve average life expectancy faster than we age—if, in other words, life expectancy increases by more than a year annually—we can hope to escape death altogether. He calls this "escape velocity."

To reach escape velocity, de Grey has a plan. Bucking the conventional wisdom of the biological community, he proposes that we can defeat aging if we crack seven key problems: (1) replenish cells that are lost or damaged over time, (2) remove senescent cells, (3) prevent stiffening of structures around the cell with age, (4) prevent mitochondrial mutations, for example by engineering mitochondria so that they don't make any proteins themselves using their own genome but import them exclusively from the rest of the cell, (5) restore the elasticity and flexibility of the structural support to cells that stiffen with age, (6) do away with telomere lengthening machinery so that we don't get cancer, and (7) figure out how to reengineer stem cells so that our cells and tissues don't atrophy. He calls his program to solve these problems SENS: strategies for engineered negligible senescence.

De Grey has learned enough biology to pinpoint many of the things that go wrong as we age. But with the characteristic arrogance that many physicists and computer scientists display toward biologists, he is wildly optimistic about the feasibility of addressing them. In response to his claims, twenty-eight leading gerontologists, including many you've come across in this book, wrote a scathing rebuttal arguing that many of his ideas were neither sufficiently well formulated nor justified to even provide a basis for debate, let alone research, and that not a single one of de Grey's proposed strategies has been shown to extend life span. The coauthors included Steven Austad and Jay Olshansky. Other mainstream researchers too dismissed SENS as pseudoscience. One of them, Richard Miller of the University of Michigan, penned a hilarious parody of SENS in a satirical open letter to de Grey in the journal *MIT Technology Review*. Since the aging problem had been solved, Miller proposed, perhaps we could turn now to the challenge of producing flying pigs; there are a mere seven reasons why pigs, at present, cannot fly, and we could fix all of them easily. De Grey, in response, huffed that the gerontology community was short-sighted, comparing the field to Lord Kelvin, the famous physicist and former president of the Royal Society who once scoffed that heavier-than-air flying machines were impossible.

Dissatisfied with the lack of support from the academic community and the funding prospects in England, de Grey left for the United States in 2009. He set up the SENS Foundation in well-heeled

Mountain View, California, with a private endowment, and initially with the support of some well-known gerontologists. Around this point, he began liaisons with other women, two of whom were forty-five and twenty-four years old. Adelaide Carpenter de Grey, then sixty-five, did not want to move to California to be part of this lifestyle, and they eventually divorced. De Grey remarked that as we solved the aging problem, "There's going to be much less difference between people of different chronological ages," and the expectation of living a very long time might very well lead to a reevaluation of the value of permanent monogamy. In 2021 he made the news again after being accused of sexual harassment by two young women, one of whom was only seventeen when she encountered de Grey. He denied the allegations and was suspended by his own foundation initially. But following charges that he'd interfered with an investigation into his conduct, the SENS Foundation fired him. A company report eventually cleared de Grey of being a sexual predator but criticized him over instances of poor judgment and boundary-crossing behavior. De Grey, undaunted, founded the new LEV Foundation, with the letters standing unsurprisingly for Longevity Escape Velocity. His longevity in longevity research is remarkable, as is his ability to continue to obtain funding from rich benefactors.

Even the more mainstream anti-aging industry has some extreme optimists. Among them is David Sinclair, who, unlike the charlatans of the aging field, is a Harvard professor who has published a number of high-profile papers on aging in top journals, including two recent papers on reprogramming cells that made considerable waves. At the same time, Sinclair is known for excessive self-promotion and highly enthusiastic claims. For example, he has predicted that it will be normal to go to a doctor and take a medicine that will make us a decade younger, and that there is no reason why we couldn't live to be 200. Such statements cause some of his critics to cringe and even fellow scientists who respect his ability to be embarrassed for him. I discussed the fate of resveratrol and his company Sirtris in chapter 8, but it appears to have had no effect on his ability to raise money to found several new companies—or indeed on his large public following, one that rivals de Grey's. His recent popular book, which doubles down on his beliefs, shows that he is completely unfazed by any criticisms of his work. I doubt whether he would have been bothered much by a scathing review of the book by Charles Brenner.

Although resveratrol has long been discounted by the mainstream community, Sinclair still stands by it. In an essay on LinkedIn, he said coyly that he does not give medical advice—then proceeded to say that he takes resveratrol, metformin, and NMN (an NAD precursor) daily. We have come across these compounds in these pages. There is no evidence that any of them improves life span in humans; they haven't been tested for this purpose in rigorous clinical trials, and, therefore, have not been approved by the FDA. Moreover, the evidence that metformin is beneficial in healthy adults is mixed; as we saw earlier, there are also problems associated with its use. For a Harvard professor to make this sort of statement on social media is essentially advocating their use, which strikes me as both ethically questionable and potentially dangerous. In the piece, Sinclair also bragged that he had a heart rate of 57 despite not being an athlete and that his lungs functioned as though he were multiple decades younger. Oddly, I am seventy-one, and although I'm no athlete either, my resting heart rate has been in the low 50s for much of my adult life—without taking Sinclair's nutraceutical supplements. Since he is a scientist, at least he ought to compare himself to close relatives who don't take the supplements, and also see what would happen if he went off his regimen but preserved his general lifestyle.

Starting a few decades ago, all sorts of dubious commercial enterprises started selling various compounds or procedures purporting to extend health or life. They would often make the most tenuous connection with some genuine research finding to hawk their wares. Respectable scientists founded their own companies— in many cases, several—and some of them gave the impression that the problem of aging would soon be solved. After all, investors are unlikely to fund companies if the payoff is many decades down the road. All of this led to a feeling that the fountain of youth was just around the corner.

Even back in 2002, fifty-one leading gerontologists were already alarmed enough by the hype to write a position statement laying out their views on what was known and what was fantasy or science fiction. They were particularly anxious to draw a clear distinction between serious anti-aging research and questionable claims about extending health and life. Among their key points:

Eliminating all aging-related causes of death would not increase life expectancy by more than fifteen years.

The prospects of humans living forever is as unlikely today as it has ever been.

Antioxidants may have some health benefits for some people, but there is no evidence that they have any effect on human aging.

Telomere shortening may play a role in limiting cellular life span, but long-lived species often have shorter telomeres than do short-lived ones, and there is no evidence that telomere shortening plays a role in determining human longevity.

Hormone supplements sold under the guise of anti-aging medicine should not be used by anyone unless they are prescribed for approved medical uses.

Caloric restriction might extend longevity in humans, since it does so in many species. But there is no study in humans that has proved it will work, since most people prefer quality of life to quantity of life; but drugs that mimic caloric restriction deserve further study.

It is not possible for individuals to grow younger, since that would require performing the impossible feat of replacing all of their cells, tissues, and organs as a means of circumventing aging processes.

While advances in cloning and stem cells may make replacement of tissues and organs possible, replacing and reprogramming the brain is more the subject of science fiction than likely science fact.

Despite these many reservations, the gerontologists enthusiastically supported research in genetic engineering, stem cells, geriatric medicine, and therapies to slow the rate of aging and postpone age-related diseases.

Interestingly, Aubrey de Grey was a signatory to this statement. Notable omissions, though, included Leonard Guarente and David Sinclair, both of sirtuin fame, and Cynthia Kenyon, who had discovered the *daf-2* mutant in worms. All three of them were involved with various longevity companies at the time and were on record as being highly optimistic about the prospects of major breakthroughs.

Nevertheless, the explosion in the anti-aging industry has

proceeded unabated. Today there are more than 700 biotech companies focused on aging and longevity, with a combined market cap of at least \$30 billion. Some of these firms have been around for almost two decades but have yet to produce a single product. Others generate revenue by selling nutraceuticals; these supplements do not require FDA approval, and no randomized clinical trials to assess their safety and effectiveness have been carried out. Many of these companies have highly distinguished scientists on their advisory boards—including some Nobel laureates who have no particular expertise in aging, apart from being old. To the public, the presence of these distinguished scientists lends an air of credibility to the enterprise. How has such an enormous industry flourished for so long with so few actual advances to show for it?

AGING RESEARCH TAPS INTO OUR primeval fear of death, with many people willing to subscribe to anything that might postpone or banish it. California tech billionaires, especially. Many of them made their money in the software industry, and because they were able to write programs to carry out rapid financial transactions or swap information of various sorts, they believe aging to be just another engineering problem to be solved by hacking the code of life. The pace of success in the software industry has made them impatient. They are used to making major breakthroughs in a couple of years, sometimes even a couple of months, and they underestimate the complexity of aging. They want to "move fast and break things." We all know how that attitude worked out for social media, with consequences for social cohesion and politics that we could never have imagined twenty years ago. Currently, these same people have prematurely unleashed AI on the world while at the same time warning us of its dangers. One can only shudder at applying that attitude to something as profound as aging and longevity.

These enthusiastic tech billionaires are mostly middle-aged men (sometimes married to younger women) who made their money very young, enjoy their lifestyles, and don't want the party to end. When they were young, they wanted to be rich, and now that they're rich, they want to be young. But youth is the one thing that they cannot instantly buy, so, not surprisingly, many of the celebrity tech billionaires—such as Elon Musk, Peter Thiel, Larry Page, Sergey Brin, Yuri Milner, Jeff Bezos, and Mark Zuckerberg—have all expressed an interest in anti-aging research. And in many cases, they are funding it. One notable exception is Bill Gates, who recognizes realistically that the best way to improve overall life expectancy remains addressing the serious health care inequalities in the world.

Recently, the company Altos Labs made a big splash, announcing a war chest of several billion dollars of investment money. It was founded by Richard Klausner and Hans Bishop with the active encouragement and financial support of Yuri Milner and several wealthy benefactors, mostly in California, reportedly including Jeff Bezos. Milner, a software billionaire originally from Russia, has had a long-standing interest in science. He founded the Breakthrough Prizes, which are among the most prestigious—and certainly the most lucrative—international awards in science. Recently, he wrote a tract titled *Eureka Manifesto: The Mission for Our Civilization*, which explains some of his thinking about aging. Some of what he believes seems to be similar to the transhumanists: our evolution of reason, and all the knowledge we humans have accumulated, is precious and should not be lost. Having Earth as our only home could be a huge risk, so we may need to populate other parts of the universe. As I read his essay, I suddenly saw why Milner would want to tackle aging. Outer space is vast, and if we have to travel hundreds if not thousands of years toward a new home, it might be nice to be able to survive the voyage. There is nothing particularly illogical about Milner's views, but they display the grandiosity—and the optimism bordering on arrogance—typical of this subset of the tech community. In any case, Altos Labs was launched with a big bang in 2022. In one swoop, the company netted some of the biggest stars in anti-aging research, luring them away from their academic positions by offering them huge resources and salaries. Altos now has campuses in both Northern and Southern California (naturally), and also in Cambridge, England, not far from my own lab.

When news of Altos Labs first leaked in the press, it was touted as a company that wanted to defeat death. Rick Klausner, its chief scientist and cochair, denied this and said that its objective is to improve healthy life span. At the launch of the Cambridge campus, he said, "Our goal is for everyone to die young—after a long time." Klausner and others also pointed out that Altos Labs offers a highly collaborative way of doing science that allows it to tackle big problems in a way that academic labs dependent on individual grants cannot. Some mentioned to me that the company hoped to be gerontology's version of Bell Labs, the famous private and commercial laboratory in New Jersey where small groups worked in highly collaborative settings to produce major breakthroughs such as the transistor, information theory, and lasers.

If tech billionaires are interested in curing aging in a hurry, many scientists are only too happy to enable them. Many truly distinguished scientists now have financial stakes in the industry, either through their own companies or as employees or consultants. This is not at all a bad thing in itself, but when I see some of them constantly touting their findings or their companies' prospects, I wonder whether they can all really believe what they are saying.

Do they not understand the complexities and difficulties ahead? Or, in the words of Upton Sinclair, is it simply that "It is difficult to get a man to understand something when his salary depends on his not understanding it"?

OF ALL THE LIVING SCIENTISTS I have described in this book, Michael Hall, who led the team that discovered TOR, is one of the most distinguished. Of aging research, he told me, "I went through a period about fifteen years ago when I was thinking a lot about TOR and aging, but was then turned off by the aging meetings I attended. They were three-ring circuses: light science and wackos walking around looking like Father Time. However, I think the field has evolved. It is now on firm ground with rigorous science."

What has changed? Mainly, gerontology has gone from being a somewhat disrespectable soft science scorned by mainstream biologists to becoming a major research priority, partly because of the need to deal with aging populations in the developed world and, increasingly, worldwide. The result is that we now have a much better handle on the complicated biological causes of aging. Of these, DNA repair, although fundamental to aging, has been used far more to target cancer than aging. Virtually every other aspect of aging is also the target of therapeutic interventions to slow it down or reverse it. We have discussed many of them in context throughout the book, but some of them seem to be more promising than others—and have certainly attracted more investment.

One promising approach is to prevent the accumulation of "bad" proteins and other molecules as we age, either by recognizing them and disposing of them, or by slowing down or altering the rate or program of protein production, which allows the body to cope with these changes. Drugs that essentially mimic caloric restriction fall into this class, and the ones that are most actively investigated are those that target TOR, such as rapamycin and similar drugs, and others like the antidiabetic drug metformin, whose mechanism of action is still not well understood. The vitamin-like precursors of NAD and other nutrients that need to be supplemented with age are also an active area of research. Other drugs aim to target senescent cells, which are the source of inflammation and its accompanying problems, while still others seek to identify factors found in young blood that can slow down aging in various ways.

Some of the biggest excitement today concerns the reprogramming of cells to reverse the effects of aging. You have already read in chapter 10 about how scientists are using transient exposure to Yamanaka factors to try to rejuvenate animals while also trying to minimize the risk of cancer. The early results of this approach have been promising enough that a huge number of start-up companies has sprouted up around this strategy. It is a major focus of Altos Labs, which hired Shinya Yamanaka himself as an adviser. Stem-cell therapy was already a major area of biotechnology because of its potential to regenerate damaged tissue and restore function to organs. Many of these companies already have expertise in reprogramming to generate various kinds of stem cells and have now jumped onto the anti-aging bandwagon. However, patients will be more receptive to stem-cell treatment for serious diseases such as replacing damaged muscle after a heart attack or restoring functional cells in a pancreas to treat diabetes, because the benefits will clearly outweigh the risks. It is not yet clear when this will happen with efforts to tackle aging—clearly the bar for safety and efficacy will be much higher.

That brings us to another, more fundamental problem with aging research. How can researchers tell if their treatments are working? The customary way for any new treatment in medicine would be to carry out a randomized clinical trial. Patients are divided into two groups, with one given either a placebo or the current standard therapy for a particular condition, and the other the agent being tested, to see if the patients given the experimental medicine fare better, or worse. The equivalent for anti-aging medicine would be to see if the treatment prolongs health and life. But this could take years to assess. This long wait for results makes it more difficult to find volunteers for properly randomized trials.

In management, as well as in science and technology, there is a well-known saying that you can't improve what you can't measure. The fifty-one gerontologists who criticized the hyperbolic statements from the anti-aging industry pointed out that aging was highly variable from individual to individual. They added pointedly: "Despite intensive study, scientists have not been able to discover reliable measures of the processes that contribute to aging. For these reasons, any claim that a person's biological or 'real age' can currently be measured, let alone modified, by any means must be regarded as entertainment, not science."

That was true twenty years ago when the authors wrote it. But today, increasingly, there are so-called biomarkers that correlate well with our underlying physiology and the characteristics that arise from it. Some characteristics of age are obvious. Our hair gets thinner and grayer or whiter, our skin becomes more wrinkled and less elastic, our arteries narrow and become more rigid, our brains are— Well, you get the picture. These traits are subjective and tricky to quantify, but if we can come up with measurable biomarkers that are proxies for them, that would be a big step forward. In addition to epigenetic changes to our DNA such as the Horvath clock, explained in chapter 5, there are now a variety of markers that measure inflammation, senescence, hormone levels, and various blood and metabolic markers, as well as the pattern of gene expression in different cell types. So scientists may be able to measure if their treatments are having any effect on aging without having to wait an interminably—or terminably—long time. Although these biomarkers or aging clocks have been rapidly taken up by the industry, their underlying basis is often not clear, and there are few studies that compare them to see how well they agree with one another.

Anti-aging researchers run into a regulatory problem as well: clinical trials are usually only approved for treatment of disease. In the scientific community, debate rages over whether aging is simply a normal progression of life or a disease. The traditional view is that something that happens to everyone and is inevitable can hardly be termed a disease. Gerontologists who subscribe to this view would argue that aging is the result of molecular changes that occur over time, which make us function less optimally and become more prone to diseases. Aging may be a *cause* of disease but is not a disease in itself. Another stark difference is that disease is usually subject to a clear definition: whether one has it and when one got it. But there is no clear consensus on when you become old. For these reasons, the latest International Classification of Diseases by the World Health Organization (WHO) omitted aging. While many in the gerontology community were disappointed by this decision, others welcomed it because they worried that classifying aging itself as a disease could lead to inadequate care from physicians: rather than pinpoint the cause of a condition, they would simply dismiss it as an unavoidable consequence of old age.

Still, the biggest risk factor for many diseases is age. Even during the recent Covid-19 pandemic, the risk of dying from being infected roughly doubled with every seven to eight years of age, so that an eighty-year-old was about 200 times as likely as a twenty-year-old to die if he or she caught Covid. Drawing on this, some gerontologists argue that we should regard aging as a disease, one that manifests itself in various ways such as diabetes, heart disease and dementia, or indeed being more prone to pneumonia or Covid-19. Of course, with billions of investment and research dollars at stake, there is currently fierce lobbying both by elements of the gerontology community and the anti-aging industry to have aging classified as a disease. So far, the FDA has refused, although it approved clinical trials for progeria, a disease in which patients age prematurely, dying around fifteen years of age. More surprisingly, in 2015 it authorized the TAME trial on the use of metformin in a study of aging in healthy adults; perhaps the federal agency was swayed by the fact that metformin was already an approved drug for diabetes, and at least some data on diabetics suggested a beneficial effect. But unless companies invested in longevity succeed in persuading the FDA to allow clinical trials for normal aging, they will face difficulty carrying out rigorous patient studies and will have to resort to other criteria to show the efficacy of their treatments.

MOST PEOPLE SAY THEY DO not fear death so much as the prolonged debilitation that precedes it. Almost everyone would agree that it is a worthy goal to increase health span, or the number of years of healthy life, by reducing the fraction of years of life that we spend in poor health as a result of age-related diseases. This goal was termed compression of morbidity by James Fries in 1980. Or as Klausner phrased it, we should all die young after a long time. Compression of morbidity rests on two assumptions: that we can alter the process of aging to postpone the onset of the diseases of aging; and that the length of life is fixed. The first, of course, is the goal of much of anti-aging research.

However, there is some debate about the second assumption. Much of the gain in life expectancy in the last hundred years was by reducing infant mortality. However, in the last few decades, tremendous advances have been made in the treatment of diseases that occur as we age, including diabetes, cardiovascular disease, and cancer. These advances have inevitably increased our life expectancy. Aubrey de Grey has argued convincingly that the gerontology community is hypocritical in rejecting life extension because treating the causes of aging will inevitably extend life and that compressing morbidity will "forever remain quixotic." Even if we accept that there is currently a natural limit of about 120 years to our life span, the reasons for that limit are not well understood beyond a vague notion that it has to do with a general breakdown of our complex biology that leads to general frailty. As de Grey points out, compression of morbidity would require us to eliminate or slow down various causes of aging, while at the same time deliberately not tackle the causes of frailty that eventually make us die. Even Steven Austad, who is far more in the mainstream of the gerontology community than de Grey, made his famous bet that advances in combating aging would enable someone currently alive to live over 150 years.

If anything, data from the Office of National Statistics in the UK suggest that rather than compressing morbidity, advances in treatment of age-related diseases have done the opposite: they show that the number of years we spend with four or more morbidities has not declined but actually slightly increased as a fraction of our lives. A United Nations report on the trend worldwide is similar and concludes that both life span and disability-free years increased but the fraction of our lives spent in disability has not decreased. In short, we are living more years and possibly a greater fraction of our lives in poor health.

Is compression of morbidity even possible? When I first heard the idea, I thought it was absurd: if someone was "young" in Klausner's sense of being healthy, what would suddenly cause him or her to collapse and die? It would be like a car that was running perfectly suddenly falling apart. In his original 1980 article on compression of morbidity, Fries himself likened the idea to the titular one-hoss-shay of the 1858 Oliver Wendell Holmes poem

"The Deacon's Masterpiece or, the Wonderful 'One-Hoss Shay'"
in which a shay—a horse-drawn carriage for one or two people—
was designed so perfectly that all its parts were equally strong and
long-lasting. A farmer was merrily riding it when all of a sudden
the shay disintegrated under him—"Just as bubbles do when they
burst"—and he found himself on the ground in a heap of dust.

There are animals that live a healthy and vigorous life, reproducing right up to the point of death. In his book *Methuselah's Zoo*,
Steven Austad describes an albatross that lives many decades in
perfect health until it dies. However, the albatross's demise is not
the death we might wish for, as centenarians in the peak of health
quietly slipping away in our sleep. In nature, life is brutish and
merciless. The bird probably reached a point where it could no
longer make the long journey to return to its nest and collapsed
after a struggle, or it was killed by a predator. Similarly, our hunter-
gatherer ancestors probably did not spend many years with the
morbidities of old age; instead, they often starved, died of disease,
were eaten by predators, or killed by a fellow human being the
moment they were not absolutely healthy and fit. Their morbidity
was highly compressed but it's not exactly what most of us are
striving for. If compressing morbidity were the only goal, we could
squish it all the way to zero if we chose. In Aldous Huxley's classic
1932 dystopian novel *Brave New World*, perfectly healthy people
are simply euthanized at their appointed time. It is not clear that
many people would opt for such a world especially if the timing of
"compression" was not up to us. If we were faced with many years
of decrepitude, some of us might well consider it, but if we were
perfectly healthy, why would we want to die? I don't think these examples represent true compression of morbidity, because the death
of an otherwise healthy being occurs rather suddenly as the result
of some unpleasant external cause.

If all this sounds bleak, there is some hope that true compression of morbidity is actually possible. Thomas Perls of the New England Centenarian Study points out that although the number of centenarians has grown in recent decades, the numbers of semisupercentenarians and supercentenarians (those that reach 105 and 110 years of age, respectively) have not and remain very small. This is contrary to what we would expect given medical advances and a general population increase in life expectancy. While many centenarians live extraordinarily long lives in good health, about 40 percent of them had age-related diseases prior to 80. By contrast, supercentenarians are healthy nearly their entire lives. As they approached the limit of the human life span at around 120 years, like the one-hoss-shay they experienced a rapid terminal decline in function and died. This would argue in favor of a fixed life span, with supercentenarians managing to compress morbidity as much as possible and pushing close to the maximum life span of the species.

Perhaps by studying their genetics, metabolism, and lifestyles, we can understand what it would take to achieve a life that is healthy right up to the very end. There may be hundreds of genetic changes that each contribute in a subtle way to longevity, and there may be no magic combination of genes that allows you to live very long. Moreover, although scientists have been able to isolate single genes that extended life in highly artificial situations, we know that those mutants are unable to compete with normal wild-type worms or flies because these genes are detrimental to fitness in other ways. Similarly, a variant of a gene called APOE is overrepresented in centenarians and is thought to protect against Alzheimer's disease, but this same variant increases the risk of metastatic cancer, and also makes people more likely to die of Covid-19. Findings like these should temper any dreams of using future advances to engineer humans with extremely long lives. Genetic variants that are associated with longevity could make us vulnerable in other unforeseen ways.

Anyway, even these supercentenarians are hardly as fit as they were in their twenties, nor indeed would you mistake them for a younger person. Something about them has still aged, and they become increasingly frail. As I pointed out earlier, Jeanne Calment was deaf and blind near the end. So the question of what characterizes good health or a lack of morbidity bears closer examination.

It is conceptually easy to define *mortality*, but *morbidity* is much fuzzier. It is defined as a disease, but many chronic illnesses such as diabetes, high-blood pressure, or atherosclerosis can be treated with medication and people can lead perfectly normal and satisfactory lives. I take medication for high cholesterol and high blood pressure, which might be termed chronic diseases, but I can do most things I like, including bicycling and hiking. If you simply count diagnoses for diseases as morbidities, then you are not capturing a true picture of whether the person is living a reasonably healthy life or is decrepit, incapacitated, and suffering. Statistics regarding morbidities in old age must be looked at carefully.

The efforts to combat aging today span a wide range. At one end are a small and highly vocal minority, including both high-profile scientists and investors, who want to defeat death altogether. They have large, cultlike followings, and I suspect there are many more who want this goal but are too embarrassed to profess it openly. At the other end are those focused strictly on treating specific diseases of old age using what we have learned about their various causes. The broad spectrum in the middle want to tackle aging directly to compress morbidity so that humans might live healthy lives into old age.

Today there is a vast amount of money invested in aging research, both by governments and by private commercial companies. In a decade or two, we will have a clear idea of whether they will succeed and to what extent. If they succeed even partly, it could have profound and unpredictable consequences for society. Let's now look at what some of those might be.

12.

SHOULD WE

LIVE FOREVER?

I am now roughly the age my grandparents were when they died. The physically active lifestyle I lead is something they could not have imagined in their final decade. Today it is increasingly common for people to die in their nineties or later. My personal experience is simply a reflection of demographic changes in the world over the last few decades. Virtually every part of the world is experiencing a growth in the size and proportion of the population over the age of sixty-five. The share of older people is currently almost 20 percent in high-income countries and expected to double between now and 2050 in many regions of the world.

At the same time, people are having fewer children. We first saw this in developed countries and are increasingly seeing it now across the globe. This means that fewer and fewer workers will support an ever larger population of retirees. In some Asian countries, there may eventually be twice as many retired people as there are workers. Many of the elderly will also require expensive medical care for a decade or even two. In countries with weak social safety nets, they will either be at the mercy of their families or will have to be self-reliant, for which they will need to be mentally and physically fit. Even in countries with more robust state support, an aging population will put tremendous strain on pension and social security programs.

The social consequences of extending life span are immense. Nearly all state-backed retirement programs assume that people will stop working around age sixty-five. These measures were introduced when people generally lived only a few years past retirement age, but now they can live two decades beyond it. In both social and economic terms, this is a ticking time bomb, and it is no surprise that governments the world over are enthusiastically funding aging research to improve health in old age in the hopes that this segment of the population can be both more productive and independent for a longer time, and in less need of costly care.

If we increase life span without compressing morbidity, it will simply make our current problems worse. But if researchers manage to combat aging *and* compress morbidity, we could well see a scenario where people routinely live healthily beyond 100 years, possibly approaching our current natural limit of about 120 years of age. In the context of any one individual that might seem a wonderful outcome, but it will also have profound and unpredictable consequences for society.

When major, disruptive technologies arrive, we are not always good at understanding their long-term ramifications. For example, not so long ago, people gladly adopted social media while giving scarcely a thought to its potential consequences, such as a loss of privacy, monetization of the individual by large corporations, surveillance by governments, and the spread of misinformation, prejudice, and hatred. We cannot afford to repeat that mistake by blindly adopting new anti-aging technologies and sleepwalking into a world for which we are ill-prepared. What might some of the consequences of life extension be?

One of them is even greater inequality. There is already a wide gap in life expectancy between the rich and poor. Even in England, which has a national health service providing universal coverage, this disparity is about ten years. However, the difference in the number of healthy years is almost twice that. The poor not only live shorter lives but also spend more of it in poor health. Things are even worse in the United States, where the richest live about fifteen years longer than the poorest, and the disparity actually increased between 2001 and 2014.

Advances in medicine have always had the potential to increase inequality. Historically, the rich in advanced countries have bene-fited first. Later, others in these countries may benefit, depending on whether health-care systems and insurance companies view these treatments as necessities. Only then will they eventually spread to the rest of the world, where only those individuals who can afford them will be able to benefit. We already see this in the health and economic status of people from different parts of the world. So any advances in aging research is likely to similarly increase inequality. But unlike other kinds of inequality, an inequality in both the quality and extent of life has the potential to be not just self-sustaining but actually to drive even larger increases in inequality. The economi-cally well off in white-collar jobs will now be able to live and work longer and pass on even more generational wealth to their descen-dants, thus exacerbating the inequality. Unless treatments become very cheap and generic—such as cholesterol-lowering statins or blood pressure medications—there is a serious risk that we will be creating two permanent classes of humans: those who enjoy much longer lives in good health, and the rest.

Another concern is overpopulation. Such a large increase in life expectancy could lead to a dramatic increase in the world's popu-lation at a time when there are already too many people on Earth. Our current population, and its predicted increase in the coming decades, is partly why we face so many existential disasters, including climate change, loss of biodiversity, and dwindling access to natural resources like fresh water.

Past increases in longevity have indeed led to dramatic increases in the population. This is because fertility rates remained high for some decades after life expectancy increased. Similarly, today, Africa has experienced significant increases in life expectancy, but fertility rates remain high at about 4.2, which is why the population of Africa is still increasing rapidly. However, improvements in life expectancy and standard of living are almost inevitably followed by a demographic transition in which the birth rate gradually falls. For example, in the late eighteenth century, European women had about five children on average at a time when life expectancy was low due to high infant mortality, but that fertility rate now ranges from 1.4 to 2.6, depending on the country. Eventually the birth and death rates became roughly equal, and the population has stabilized at some new higher level. Over the course of the nineteenth and twentieth centuries, this happened in much of the West, as well as in many Asian countries such as Japan and South Korea.

In the past, improvements in infant and childhood mortality meant more people lived to reach reproductive age, which naturally led to rapid population growth. But it is not inevitable that in advanced countries that have already gone through a demographic transition, further increases in life expectancy will necessarily lead to a growth in population. In Japan, people live longer than they did a few decades ago, yet the population of Japan has actually fallen since 2010, because of lower birth rates.

The fertility rate has dropped and is below replacement level in many countries. The average age of childbearing has also been steadily increasing in developed countries. Currently, it is increasingly common for women to have their first child in their thirties, and sometimes even around forty, which is almost a decade or two later than the norms a century ago. Both of these trends are the result of more security and prosperity, the expectation of a long life, and the emancipation of women and their entry into the workforce. Together these factors have slowed down or stopped population growth in many parts of the world, which has been hugely beneficial in many important ways, not least the effect on our environment and natural world. I am puzzled by economists who talk about it as a problem, especially in reference to China's decline in population growth. Elon Musk believes that an impending global population collapse is a much bigger problem than climate change, which strikes me as absurd.

Nevertheless, as people live longer, the population will grow unless one of two things happens: either the fertility rate decreases even more, or the average age of childbearing increases along with life expectancy. However, both of these scenarios have some problems. In many countries, the average age of childbirth has gradually increased until it is pushing up against the realities of biology. Women from their midthirties on have increasing difficulty in conceiving and soon afterward face menopause. If menopause can be delayed as we increase life expectancy, this would solve the problem of delaying childbirth and would be much fairer to women, many of whom face the problem of deciding whether to have children right when their career is taking off. However, menopause is the result of very complex biology, and there is no evidence that we will be able to alter the age of its onset. Of course, there are ways for women to have children even beyond menopause— for example, by freezing eggs for later implantation along with hormone treatment—but these are expensive and cumbersome, and not without considerable risk. The other solution to prevent population growth in the face of increasing longevity is to have even fewer children, which means that an even greater proportion of the population will be elderly, which has its own consequences.

Let us assume an optimistic scenario: life expectancy surges beyond a hundred years and they are mostly healthy years. The population has stabilized; people are having fewer children and having them as late as possible. If we can't ask a smaller and smaller fraction of younger people to support an increasing cohort of older people in retirement, there's really only one solution: careers are going to get longer.

WORKING INTO YOUR SEVENTIES OR eighties—or even longer—is a rather different prospect depending on what your job is. As Paul Root Wolpe, director of the Emory University Center for Ethics, asks: Would hard laborers or people doing menial jobs at the age of sixty-five relish the prospect of doing this for another fifty years? Large percentages of people dislike their jobs and look forward to retirement. In 2023 more than 1.2 million people marched in France to protest against the government's proposal to raise the retirement age a mere two years from sixty-two to sixty-four. Reacting to the French protests, some have argued that the United States should actually *lower* retirement age, pointing out that the people who advocate that Americans should work until they are seventy are typically in cushy, remunerative white-collar jobs that are fun and intellectually engaging for octogenarians, and it is different for people who want to stop changing tires or working a cash register for $11 an hour at age sixty-two. In my own institute, I have found that nonscientists on the staff retire as soon as they qualify, while the scientists try to hang on for as long as they can.

When I ask some of my scientific colleagues about their retirement plans, especially in America, where it is not uncommon to see academics work well into their eighties or even longer, the typical response is "I'm having far too much fun to retire!" Some of them go on to claim they are doing the best work of their lives.

But the evidence says otherwise. We are all willing to accept that we cannot run a hundred-meter race as fast as we could when we were twenty, but we persist in the delusion that we are intellectually just as capable as we were when we were younger. This may be because we identify too closely with our own thoughts—they define who we are. All the evidence suggests that in general, we are no longer as creative and bold as when we were younger.

One way to assess this is to retrospectively ask how old someone was when they did their best work. In the sciences, Nobel Prize winners nearly always make their key breakthroughs when they are young and not very powerful. Biologists and chemists often achieve their big breakthroughs a decade or so later than physicists and mathematicians, perhaps because it takes time to assimilate a huge body of knowledge, acquire the practical experience, and build up the resources needed. Indeed, the famous mathematician G. H. Hardy wrote in his 1940 book, *A Mathematician's Apology*, "No mathematician should ever allow himself to forget that mathematics, more than any other art or science, is a young man's game. . . . I do not know of an instance of a major mathematical advance initiated by a man past fifty." In recent times, one of the great achievements of mathematics, the proof of the 350-year-old Fermat's Last Theorem, was made by Andrew Wiles when he was about forty.

When they are older, many scientists continue to churn out first-rate work from their labs. However, this is not because they themselves are sharp and innovative. Rather, they have become a brand name, have amassed resources and funding, and can attract first-rate young scientists to do the work. Many, if not all, of the new ideas—and certainly the lion's share of the work—come from these young scientists. Even so, it is very rare for an older scientist— even one who is doing very good work and has a team of young scientists to help—to truly break new ground. Often they are doing more of the same. For example, I have had the good fortune to attract very talented young people thanks to whom my laboratory continues to publish papers in top journals. But it is also true that in some sense, they are extensions of my previous work. The few really new directions have come not from me but from the young people who work with me. It is true that everyone can point to an exception: the chemist Karl Sharpless won his second Nobel Prize at the age of eighty-one for work he had begun when he was around sixty. But that is remarkable because it is so rare.

It is not just in science and mathematics that our creative powers peak when we are relatively young. This is also true in business and industry. Thomas Edison was under thirty when he started the Menlo Park laboratory in New Jersey and invented his version of the lightbulb soon afterward. In today's world, many of the most innovative companies, such as Google, Apple, Microsoft, and the AI company DeepMind, were started by people in their twenties or thirties.

You might think that things are different in literature, where experience of life and accumulated wisdom would make you more profound as you aged. However, at a Hay Literary Festival event in 2005, the Nobel Prize–winning novelist Kazuo Ishiguro outraged his fellow writers by suggesting that most authors produce their best work when they are young. He said it was hard to find cases where an author's most renowned work had come after the age of forty-five and pointed out that *War and Peace, Ulysses, Bleak House, Pride and Prejudice, Wuthering Heights*, and *The Trial* were all written by writers in their twenties and thirties. Many great writers— Chekhov, Kafka, Jane Austen, the Brontë sisters—died before they reached their midforties. Ishiguro says he is not suggesting that novelists cannot do good work later in life, just that their best work tends to come before their midforties. His main point was actually that authors should not wait until they are older to attempt a great novel. He may have contradicted his own thesis with *Klara and the Sun*, which he wrote in his midsixties. It was received as one of his finer novels, although only time will tell whether it will rank as highly as his earlier work. Similarly, Margaret Atwood's recent Booker Prize–winning novel, *The Testaments*, was published when she was over eighty. It is brilliantly gripping and disturbing, but the novel is really a further exploration of the world she conjured in *The Handmaid's Tale* almost forty years before.

Ishiguro posited a theory for why some types of creativity decline with age. As we grow older, one of the first mental abilities to decline is our short-term memory. Perhaps writing a novel requires holding disparate facts and ideas in our heads while we synthesize something new from them. This may well be true in science and mathematics. The process of creativity may be different in other disciplines. For example, many film directors, conductors, and musicians continue to perform at the highest level well into old age, as do many artists.

Advances in healthy aging would not necessarily make us as creative and imaginative later in life as we are in our younger years. Young people see the world with fresh eyes, and in new ways. Ishiguro wonders whether in writing, the proximity to childhood and the experiences of growing up—a time of life when one's perspective changed from year to year, even month to month, because one was oneself changing so profoundly—is central to the creation of satisfying novels. In science and mathematics, younger practitioners may be less biased by a lifetime accumulation of knowledge, and bolder about questioning paradigms.

So far, we have been talking about big creative breakthroughs declining with age in a variety of fields, but these breakthroughs are outliers and represent a tiny fraction of the whole enterprise. Even in science, the big breakthroughs are built on the vast foundations laid by the majority of scientists productively going about their jobs of gradually advancing our state of knowledge. It would hardly be appropriate to formulate social policy based on these outliers. How would the bulk of white-collar work be affected by age?

Most studies say our general cognitive abilities also decline with age, but there has been some debate about when exactly that happens, with some arguing that it begins as early as age eighteen, and others arguing that it is significant only after sixty. A ten-year study that followed a large cohort of British civil-service workers showed that cognitive scores on tests of memory, reasoning, and verbal fluency all declined from the age of forty-five, with faster decline in older people. The one category not to show a major decline was vocabulary. Other studies also make a distinction between so-called "crystallized abilities" such as vocabulary and "fluid abilities" such as processing speed. The latter declines steadily from the age of twenty, while the former increases and then remains steady, and only declines gradually from about age sixty. All of this affects our ability to learn new tasks and be as mentally agile. Any adult who doubts these findings should try learning the piano, a new language, or advanced mathematics for the first time.

It is of course theoretically possible that as we learn to combat the causes of aging, we can also do something about the deterioration of our mental abilities. But so far, the brain has proved the most difficult frontier to conquer. Neurons regenerate very slowly if at all, and many of the processes that lead to deterioration and eventual disease in the brain remain intractable. It is true that at least one approach, inhibiting the integrative stress response in protein synthesis, has been shown to improve memory, but there is no evidence that it reverses general cognitive decline and ability to learn.

Many argue that any cognitive decline is offset by increased wisdom, a vague and poorly defined trait. It's true that young people often do lack wisdom and foresight, leading to rash behavior. But there is no evidence that wisdom continues to increase beyond a certain age. In recent elections in both the United States and

Great Britain, older age groups have tended to be conservative and swayed by demagoguery and an appeal to their sense of nostalgia. They have acquired a lifetime of biases and prejudices and are generally less open to new ideas. My guess is that we acquire most of our wisdom by our thirties. After that, we become increasingly set in our ways, as likely to be reactionary as wise.

Today there is an imbalance of power that favors the old. This is partly because they have accumulated a great deal of wealth: in both Britain and American, households where the head is over seventy have about fifteen to twenty times the median wealth of those under thirty-five. But it is also because as people age, they accumulate power and a powerful network of connections. Even if they are no longer as qualified or competent to do their job as their younger peers might be, they may cling to power and authority, using their connections and reputation. It is hard to dislodge them from their positions even if they are no longer on top of their game and could be replaced by many more competent people. More generally, Wolpe argues that the political ramifications of a long life span are huge because the elderly vote at much higher rates than the young, and the highest echelons of power have become the preserve of the over-seventies. The United States is led by President Joe Biden, who will be eighty-one as of the 2024 presidential election; his chief rival, Republican Donald Trump, will be seventy-eight. Elsewhere, Rupert Murdoch, until recently the chair of Fox Corporation and executive chairman of News Corp, retains enormous media influence (and with it, political clout) in several countries at the age of ninety-three. Politically, Wolpe argues, young people will be squeezed out, and the fresh ideas they bring to politics and innovation will be suppressed. By contrast, the vast majority of the great innovations, including social advances such as gay marriage, diversity inclusion movements, and before that civil rights and women's rights, were driven by young people.

The imbalance of power is particularly egregious in academia, where the concept of tenure, which was introduced so faculty members could not be fired for expressing unorthodox opinions, is now being wielded by faculty members to remain in their posts for as long as they possibly can. Many universities in the United States and United Kingdom have abolished mandatory retirement age, and those that haven't, such as Oxford and Cambridge, are facing lawsuits from disgruntled professors. Recently, Oxford lost a tribunal case brought by three professors who accused the university of ageism, claiming, not surprisingly, that they were dismissed "at the peak of their careers."

Even if they are not doing groundbreaking work or at the peak of their careers, as long as they are being productive, what harm is there in allowing them to stay on? Some of my academic colleagues argue that established senior scientists have the resources, wisdom, vision, and perspective to provide a great environment to train and mentor the next generation of younger scientists. Not everyone agrees. Fred Sanger, who won two Nobel Prizes, hung up his hat the day he turned sixty-five and spent the rest of his life pursuing hobbies such as building a boat that he sailed around Britain and growing roses. My own mentor, Peter Moore, retired after a long and distinguished career at Yale at the age of seventy. It is not as if he suddenly became intellectually dead. He continues to edit journals, write books, and carry on other intellectual activities that take neither resources nor money from his institution. He had this to say: "I had been telling my colleagues for years that it is an abuse of the privilege of tenure for elderly faculty to hang on to the bitter end, not least because there are no seventy-year-old scientists so wonderful that a thirty-five-year-old scientist who is better cannot be found."

In academia, the combination of tenure and a lack of retirement age is particularly problematic. Some senior academics have rightly complained that they are far more productive than some younger faculty who have burned out by the age of forty. But this can be solved by abolishing both tenure and retirement age and having regular assessments of productivity.

Moore's comment goes to the heart of intergenerational fairness. The most senior faculty tend to draw very large salaries, which would often be sufficient to hire two young scientists in their stead. Even if they are not drawing a salary, they are taking up precious resources such as laboratory space that could otherwise be used to recruit new young faculty who would go on to make the breakthroughs of the future and open up entirely new areas. Older researchers also have the clout to influence the agenda at their institution and in science more generally, and tend to be conservative and incremental rather than bold and innovative. The same is true broadly in other sectors of work, including corporate careers.

The problem of intergenerational fairness conflicts with the push for people to work longer as the population ages. So what is to be done?

Ageism is now considered a sin along with other -isms such as racism and sexism. However, ageism is different because we all actually decline with age. Still, it is important to recognize that the rate at which people's physical and mental abilities decline is highly variable. We must not use chronological age as a proxy for ability, and a rigid retirement age that applies to everyone is highly inappropriate. Moreover, despite the well-documented decline in people's ability with age, two surveys of the literature concluded that the relationship between age and productivity is more complex. One concluded that as they aged, people did less well at tasks that required problem-solving, learning, and speed, but maintained high productivity in jobs where experience and verbal abilities are important. The other concluded that 41 percent of the reports showed no differences between younger and older workers, and

28 percent reported that older workers had better productivity than younger workers, citing experience and emotional maturity as possible factors.

All of this suggests that we need to be flexible in our approach to work and retirement. As we have seen, many professions are physically or mentally demanding, and people may need to retire earlier. They may be able to switch to less demanding jobs and continue working if they are able. Rather than apply a one-size-fits-all approach, we need to bring in objective measures of assessment that can apply to all age groups, which will also ensure fairness to both young and old. Moreover, even after they can no longer do the job they did for much of their career and have to retire, older people can still be useful and productive in many ways for as much of the rest of their lives as possible.

There is a lot of evidence that having a purpose in life reduces mortality from all causes as well as the incidence of stroke, heart disease, mild cognitive decline, and Alzheimer's. And elderly professionals do have a wealth of experience and a deep knowledge of their field. They can be unparalleled sources of advice and mentorship; they can participate in civic activities. Peter Moore, whom I mentioned earlier, is a great example of someone who has retired from his professorship but still makes himself extremely valuable to the scientific community.

Even after they have retired, we need to think of ways that allow older citizens to remain independent for as long as possible. This means paying attention to the way houses are constructed, with bedrooms on ground floors, and communities are planned, with nearby amenities such as shopping and mass transit. Social isolation and loneliness are detrimental for the well-being of all people but especially for the elderly. Currently, many Western societies seem to treat the old as a problem to be hidden away in separate retirement enclaves rather than an integral part of society.

Perhaps it is better to integrate them fully into the broader community, where they live interspersed with the rest of the population, and through their social and civic activities, they interact routinely and regularly across the entire generational spectrum of society. Their active participation will also benefit the rest of society.

These are all problems we may plausibly soon encounter, if biologists succeed in pushing life spans ever closer to a natural limit of roughly 120 years. Yet there is no hard scientific law that necessarily precludes far more drastic increases in life expectancy. After all, we know of species that live many hundreds of years and others that show no signs of biological aging. If, someday, humans breach our current limit and live for several hundred years as Aubrey de Grey prophecies, all of these issues would only be magnified. Advocates for extreme life extension have no real solutions except to say that we will learn to deal with problems as we encounter them. Some have said that if we have a population crisis as a result of extreme longevity, we should be made to leave Earth and settle other planets once we reach a certain age. As always, the answer to problems created by technology seems to be even more far-fetched technology.

I AM NOT SURE THAT if we lived so much longer, we would be any more satisfied. Now that we live twice as long as we did a century ago, we still aren't content with that entire extra life. Rather, we seem to be even more obsessed with death. If we live to be 120 or 150 years old, we will fret about why we can't live to 300. The quest for life extension is like chasing a mirage: nothing will ever be enough short of true immortality. And there is no such thing. Even if we conquer aging, we will die of accidents, wars, viral pandemics, or environmental catastrophes. It may be simpler to accept that our life is limited.

Moreover, our very mortality may give us the incentive and desire to make the most of our time on Earth. A greatly extended life span would deprive our lives of urgency and meaning, a desire to make each day count. It is not clear that even with an entire extra lifetime, we are accomplishing more than the great writers, composers, artists, and scientists of past eras. We may well end up living a very much longer life bored and lacking in purpose. As I mentioned earlier, it could also lead to a stagnant society, since many of the big social changes have been spearheaded by younger generations.

This obsession with mortality is probably unique to humans. It is only the accidental evolution of our brain and consciousness, and our development of language to communicate our fears, that has made our species so fixated on the end. The writer and editor Allison Arieff has pointed out the irony that the same Silicon Valley culture that produces gadgets designed to be obsolete and discarded every few years seems to be obsessed with living forever. She quotes the writer Barbara Ehrenreich, "You can think of death bitterly or with resignation and take every possible measure to postpone it. Or, more realistically, you can think of life as an interruption of an eternity of personal nonexistence, and seize it as a brief opportunity to observe and interact with the living, ever-surprising world around us." Arieff believes that our very humanness is intertwined with the fact of our mortality.

On a recent trip to India, I met Ganesh Devy, a linguist who works with dozens of rural, forest-dwelling tribes in the country. India has well over a hundred languages, many facing a different kind of death: some of them are now spoken by only a few people and will soon become extinct. He said he himself did not fear death. I was skeptical, but he pointed out that on a field trip once he was bitten by a highly poisonous snake and he felt no fear or panic at the thought of dying. I asked him why. Devy said that we have to regard our individual selves as parts of larger entities like family, community, and society, just as all the cells in our body are part of tissues and organs and us. Millions of our cells die every day. Not only do we not mourn their passing, but we are not even aware of it. So even if we as individuals die, our society and indeed life on Earth will go on. Our own genes will live on through our offspring or other family members. Life has been going on continuously for several billion years while we individuals come and go.

Still, if someone were to offer a pill that would add ten years of healthy life, hardly anyone would decline it. I view myself as more in the philosophical camp, yet take several anti-aging medicines a day: pills for my blood pressure, a statin for high cholesterol, and a low-dose aspirin to protect against thrombosis. All of these are to prevent heart attacks or strokes and have the effect of prolonging my life. I would be a hypocrite to dismiss attempts to alleviate the problems of aging. Physicians are struck by how many people, even faced with terminal illnesses that inflict appalling pain, want every measure taken to prolong their lives, even if only by a few weeks or even days. The will to live is deeply ingrained in us, even if we are sanguine in our more rational moments.

About ten years ago, the Pew Research Center explored American attitudes on living much longer. Respondents were optimistic about cures for cancer and artificial limbs, and they viewed advances that prolong life as generally good. However, over half said that slowing the aging process would be bad for society. When asked if they themselves would take treatments to live longer, a majority of them said no, but two-thirds thought that *other* people would. Most doubted that an average person living to 120 would happen before 2050. A large majority felt that everyone should be able to get these treatments if they wanted, but two-thirds felt that only the wealthy would actually have access. About two-thirds also said that longer lives would strain our natural resources. About six in ten said that medical scientists would offer treatments before they fully understood how doing so could affect people's health and that such treatments would be fundamentally unnatural. The clear-eyed view of the American public in the face of relentless hype is certainly heartening.

In this book, I have discussed how advances in molecular biology have shed light on virtually every aspect of aging, often taking a skeptical look at some of the hype. In doing so, I hope that readers acquire not only an appreciation of the underlying causes of aging, but are able to more knowledgeably interpret news reports and PR blurbs about each new "advance" and judge for themselves how realistic various claims are. How long it takes to go from a fundamental discovery to a practical application is hugely variable and unpredictable. It took three centuries for Newton's laws of motion to be translated into rockets and satellites. It took over a hundred years for Einstein's theories of relativity to be used in the GPS systems that our phones use to tell us where we are on a map. Neither Newton nor Einstein could have remotely anticipated the use we made of their discoveries. Other advances are much faster: from Alexander Fleming's discovery of penicillin in 1928 to its use in humans was less than twenty years. With the money and urgency that drive current research on aging, major advances might well come in years rather than decades, but the sheer complexity of aging makes any prediction highly uncertain.

We are at a crossroads. The revolution in biology continues unabated. Artificial intelligence and computing, physics, chemistry, and engineering are all being brought to bear on what was the domain of traditional biologists. Together they are creating new technologies and increasingly sophisticated tools to manipulate cells and genes to advance every aspect of the life sciences, including aging.

I have highlighted the relationship between cancer and aging many times throughout this book. Both are rooted in highly complex biology. Just as cancer is not a single disease, aging too has many interconnected causes. It has now been half a century since President Nixon declared a "war on cancer" in 1971. Since then, our biological understanding of cancer has advanced enormously, resulting in a steady stream of new and improved treatments that continues to this day, saving or prolonging millions of lives. Today, the sheer talent and money committed to aging research is reminiscent of our efforts to combat cancer. This means that just as with cancer, we will eventually make breakthroughs, even if it takes time for them to actually improve and extend our lives. It is well to remember that even today, after a half century of intense effort, cancer is not "solved." It remains one of the largest killers in most societies. Our progress with aging may follow a similar trajectory, given the similar complexity of both problems.

The American futurist and scientist Roy Amara said that we tend to overestimate the effect of a technology in the short run and underestimate its effect in the long run. This has been true for many things, including the internet and artificial intelligence. If Amara's law holds, all the hype in the anti-aging industry will lead to considerable disappointment in the short term, but it also means that once we get past the winter of disillusionment and discontent, there will be major advances eventually.

As a society, it is important for us to think about the possibly profound consequences of these changes. However, this task is not just for governments and citizens alone: the anti-aging industry should not repeat the mistakes of the computer industry and plunge ahead without any thought of where it will all lead and leave the rest of us to try and clean up the mess when it is too late. These companies stand to benefit hugely from any breakthroughs in aging research but do not seem to have put much effort into either the social or ethical consequences of their work. In their blurbs, their work is always portrayed as an unmitigated and universal good for humanity.

In the meantime, we need not sit around and wait for a long period of decrepitude and decline. Ironically, the very same advances in biology that are the basis of the anti-aging industry also thoroughly validate some age-old advice for living a long and healthy life: diet, exercise, and sleep. In his book *In Defense of Food: An Eater's Manifesto*, Michael Pollan advises us, "Eat food. Not too much. Mostly plants." This advice is entirely consistent with everything we know about caloric restriction pathways. Exercise and sleep, as we discussed earlier, affect a large number of factors in aging, including our insulin sensitivity, muscle mass, mitochondrial function, blood pressure, stress, and the risk of dementia. These remedies currently work better than any anti-aging medicine on the market, cost nothing, and have no side-effects.

While we wait for the vast gerontology enterprise to solve the problem of death, we can enjoy life in all its beauty. When our time comes, we can go into the sunset with good grace, knowing that we were fortunate to have taken part in that eternal banquet.

ACKNOWLEDGEMENTS

This book would not have been written without the encouragement of Max and John Brockman, who were supportive right from the time it was merely an idea, and helped shape it into an outline that could be turned into a book.

I thank my editors, Kirty Topiwala of Hodder and Stoughton and Nick Amphlett of William Morrow, for their faith in commissioning this book and for their superb editing. Their thoroughness, insights, frank criticism, and feedback helped turn my early, impenetrable drafts into a readable narrative. I also thank Philip Bashe for the depth and thoroughness of his copyediting, which greatly improved the final text.

I owe a debt of gratitude to many scientists who freely gave of their time to offer me their opinion and feedback. First and foremost, I am grateful to Linda Partridge, one of the world's leading aging researchers, who not only helped me navigate the enormous landscape of the aging research literature, but also indulged my many questions throughout the process, and read the entire first draft of the book to give me her critical feedback. I also had many discussions with Julian Sale, Ketan Patel, and Manu Hegde, all of whom also read and provided useful feedback on the first draft. Michael Hall, Stephen O'Rahilly, and Raul Mostovslasky gave me useful comments on the material in chapters 7 and 8, and Michel Goedert in chapter 6.

David Krakauer and Geoffrey West of the Santa Fe Institute, and participants in their workshops on aging, helped me think more broadly about human aging as a special case of the growth and decline of complex systems.

In addition, a large number of people talked to me about various topics covered in the book, often pointing me to relevant findings and clarifying confusing points. They include Madan Babu, Annette Baudisch, Anne Bertolotti, Maria Blasco, Stephen Cave, Julie Cooper, Tom Dever, Alan Hinnebusch, Matt Kaeberlein, Brian Kennedy, Tom Kirkwood, Titia de Lange, Nils-Göran Larsson, Trudy Mackay, Andrew Nahum, Tom Perls, Lalita Ramakrishnan, Wolf Reik, David Ron, Melina Schuh, Manuel Serrano, Marta Shahbazi, Azim Surani, Mark Troll, Alex Whitworth, and Roger Williams. In addition, numerous others answered specific queries.

Given the state of the field, the views of the people I spoke with were not always in complete accord. In the end, I am solely responsible for the choice of material and the opinions expressed in this book.

The aging literature is vast, as is the number of researchers in the field. It would be impossible to mention them all in a book of this scope, so I have necessarily had to pick and choose in order to weave a narrative around the topic.

NOTES

Introduction

1   **Even Carter, a seasoned Egyptologist:** Maite Mascort, "Close Call: How Howard Carter Almost Missed King Tut's Tomb," *National Geographic* online, last modified March 4, 2018, https://www.nationalgeographic.com/history/magazine/2018/03-04/findingkingtutstomb.

1   **We may be tempted to think of it:** Nuria Castellano, "The Book of the Dead Was Egyptians' Inside Guide to the Underworld," *National Geographic* online, last modified February 8, 2019; Tom Holland, "The Egyptian Book of the Dead at the British Museum," *Guardian* online, last modified November 6, 2019, https://www.theguardian.com/culture/2010/nov/06/egyptian-book-of-dead-tom-holland.

2   **They recognize when one:** For example, see this study of elephants: S. S. Pokharel, N. Sharma, and R. Sukumar, "Viewing the Rare Through Public Lenses: Insights into Dead Calf Carrying and Other Thanatological Responses in Asian Elephants Using YouTube Videos," *Royal Society Open Science* 9, no. 5 (May 2022), https://doi.org/10.1098/rsos.211740, described in Elizabeth Preston, "Elephants in Mourning Spotted on YouTube by Scientists," *New York Times* online, May 17, 2022, https://www.nytimes.com/2022/05/17/science/elephants-mourning-grief.html.

2   **But there is no evidence:** James R. Anderson, "Responses to Death and Dying: Primates and Other Mammals," *Primates* 61 (2020): 1–7; Marc Bekoff, "What Do Animals Know and Feel About Death and Dying?," *Psychology Today* online, last modified February 24, 2020, https://www.psychologytoday.com/gb/blog/animal-emotions/202002/what-do-animals-know-and-feel-about-death-and-dying.

2   **Philosopher Stephen Cave argues:** Stephen Cave, *Immortality: The Quest to Live Forever and How It Drives Civilization* (New York: Crown, 2012).

3   **The first emperor of a unified China:** Ibid.

4 **Rather, our brains appear:** Y. Dor-Ziderman, A. Lutz, and A. Goldstein, "Prediction-Based Neural Mechanisms for Shielding the Self from Existential Threat," *NeuroImage* 202 (November 15, 2019): art. 116080, https://doi.org/10.1016/j.neuroimage.2019.116080, cited in Ian Sample, "Doubting Death: How Our Brains Shield Us from Mortal Truth," *Guardian* online, last modified October 19, 2019, https://www.theguardian.com/science/2019/oct/19/doubting-death -how-our-brains-shield-us-from-mortal-truth.

1. The Immortal Gene and the Disposable Body

11 **But it turns out to be tricky:** A group at the Santa Fe Institute led by David Krakauer and Geoffrey West has held several workshops to define both death as it applies to various entities and the definition of the individual.

12 **The loss of brain function:** A meeting about the issue of resuscitation and death was held at the New York Academy of Sciences in 2019. See "What Happens When We Die? Insights from Resuscitation Science" (symposium, New York Academy of Sciences, New York, November 18, 2019), https://www.nyas.org/events/2019/what -happens-when-we-die-insights-from-resuscitation-science/. There is also a movement to make the definition of brain death uniform to prevent legal anomalies such as the one I described.

13 **Her family petitioned:** S. Biel and J. Durrant, "Controversies in Brain Death Declaration: Legal and Ethical Implications in the ICU," *Current Treatment Options in Neurology* 22, no. 4 (2020): 12, https://doi.org/10.1007/s11940-020-0618-6.

13 **After that, there is a multiday window:** Two popular books that discuss these early events are Magdalena Zernicka-Goetz and Roger Highfield, *The Dance of Life: The New Science of How a Single Cell Becomes a Human Being* (New York: Basic Books, 2020), and Daniel M. Davis, *The Secret Body: How the New Science of the Human Body Is Changing the Way We Live* (London: Bodley Head, 2021).

13 **Death can occur at every scale:** Geoffrey West, *Scale: The Universal Laws of Growth, Innovation, Sustainability, and the Pace of Life in Organisms, Cities, Economies, and Companies* (New York: Penguin Press, 2020).

17 **However, the lecture paved:** R. England, "Natural Selection Before the *Origin*: Public Reactions of Some Naturalists to the Darwin-Wallace Papers," *Journal of the History of Biology* 30 (June 1997): 267–90, https://doi.org/10.1023/a:1004287720654.

17   **Although humans have known:** Matthew Cobb, *The Egg and Sperm Race: The Seventeenth-Century Scientists Who Unlocked the Secret of Sex, Life and Growth* (London: Simon & Schuster, 2007).

17   **The germ-line cells, protected in the gonads:** Today we know that the Weismann barrier is not perfect and that the germ line also ages and is susceptible to changes from the environment, although much more slowly. P. Monaghan and N. B. Metcalfe, "The Deteriorating Soma and the Indispensable Germline: Gamete Senescence and Offspring Fitness," *Proceedings of the Royal Society B (Biological Sciences)* 286, no. 1917 (December 18, 2019): art. 20192187, https://doi.org/10.1098/rspb.2019.2187.

18   **"Nothing in biology makes sense":** T. Dobzhansky, "Nothing in Biology Makes Sense Except in the Light of Evolution," *American Biology Teacher* 35, no. 3 (March 1973): 125–29, https://doi.org/10.2307/4444260.

18   **If an individual had a mutation:** T. B. Kirkwood, "Understanding the Odd Science of Aging," *Cell* 120, no. 4 (February 25, 2005): 437–47, https://doi.org/10.1016/j.cell.2005.01.027; T. Kirkwood and S. Melov, "On the Programmed/Non-Programmed Nature of Ageing Within the Life History," *Current Biology* 21 (September 27, 2011): R701–R707, https://doi.org/10.1016/j.cub.2011.07.020. There are some exceptions to this rule against group selection, but they apply only under very special circumstances and usually involve species where the members of the colonies are all genetically either identical or very closely related, such as insects. J. Maynard Smith, "Group Selection and Kin Selection," *Nature* 201 (March 14, 1964): 1145–47, https://doi.org/10.1038/2011145a0.

19   **Species such as the soil worm:** Species that reproduce multiple times in a lifetime are called iteroparous, and those that reproduce only once are semelparous. See T. P. Young, "Semelparity and Iteroparity," *Nature Education Knowledge* 3, no. 10 (2010): 2, https://www.nature.com/scitable/knowledge/library/semelparity-and-iteroparity-13260334/.

19   **He was a socialist:** N. W. Pirie, "John Burdon Sanderson Haldane, 1892–1964," *Biographical Memoirs of Fellows of the Royal Society* 12 (November 1966): 218–49, https://doi.org/10.1098/rsbm.1966.0010; C. P. Blacker, "JBS Haldane on Eugenics," *Eugenics Review* 44, no. 3 October (1952): 146–51, https://www.ncbi.nlm.nih.gov/pmc/articles/PMC2973346/.

20   **A stained glass window:** Two opposing views of Fisher can be found in A. Rutherford, "Race, Eugenics, and the Canceling of

Great Scientists," *American Journal of Physical Anthropology* 175, no. 2 (June 2021): 448–52, https://doi.org/10.1002/ajpa.24192, and W. Bodmer et al., "The Outstanding Scientist, R. A. Fisher: His Views on Eugenics and Race," *Heredity* 126 (April 2021): 565–76, https://doi.org/10.1038/s41437-020-00394-6.

20   **However, the same could not be said:** T. Flatt and L. Partridge, "Horizons in the Evolution of Aging," *BMC Biology* 16 (2018): art. 93, https://doi.org/10.1186/s12915-018-0562-z.

20   **That understanding came when British biologist Peter Medawar:** N. A. Mitchison, "Peter Brian Medawar, 28 February 1915–2 October 1987," *Biographical Memoirs of Fellows of the Royal Society* 35 (March 1990): 281–301, https://doi.org/10.1098/rsbm.1990.0013.

21   **Similarly, the disposable soma hypothesis:** Kirkwood, "Understanding the Odd Science of Aging," 437–47, https://doi.org/10.1016/j.cell.2005.01.027.

21   **Exactly as these theories would predict:** Flatt and Partridge, "Horizons," https://doi.org/10.1186/s12915-018-0562-z.

22   **But an unusual analysis:** R. G. Westendorp and T. B. Kirkwood, "Human Longevity at the Cost of Reproductive Success," *Nature* 396 (December 24, 1998): 743–46, https://doi.org/10.1038/25519. See also the letter responding to this article: D. E. Promislow, "Longevity and the Barren Aristocrat," *Nature* 396 (December 24, 1998): 719–20, https://doi.org/10.1038/25440.

23   **Menopause may have arisen:** G. C. Williams, "Pleiotropy, Natural Selection and the Evolution of Senescence," *Evolution* 11, no. 4 (December 1957): 398–411.

23   **For example, although the fertility of elephants:** M. Lahdenperä, K. U. Mar, and V. Lummaa, "Reproductive Cessation and Post-Reproductive Lifespan in Asian Elephants and Pre-Industrial Humans," *Frontiers in Zoology* 11 (2014): art. 54, https://doi.org/10.1186/s12983-014-0054-0.

23   **Similarly, while living beyond:** J. G. Herndon et al., "Menopause Occurs Late in Life in the Captive Chimpanzee (*Pan Troglodytes*)," *AGE* 34 (October 2012): 1145–56, https://doi.org/10.1007/s11357-011-9351-0.

23   **The grandmother hypothesis:** K. Hawkes, "Grandmothers and the Evolution of Human Longevity," *American Journal of Human Biology* 15, no. 3 (May/June 2003): 380–400, https://doi.org/10.1002/ajhb.10156; P. S. Kim, J. S. McQueen, and K. Hawkes, "Why Does Women's Fertility End in Mid-Life? Grandmothering and Age at

Last Birth," *Journal of Theoretical Biology* 461 (January 14, 2019): 84–91, https://doi.org/10.1016/j.jtbi.2018.10.035.

24 **Another idea, based on studying killer whales:** D. P. Croft et al., "Reproductive Conflict and the Evolution of Menopause in Killer Whales," *Current Biology* 27, no. 2 (January 23, 2017): 298–304, https://doi.org/10.1016/j.cub.2016.12.015.

24 **It could also simply be that the number of eggs:** An idea suggested to me by the population biologist Trudy Mackay of Clemson University.

24 **So perhaps there has just not been enough time:** Steven Austad, *Methuselah's Zoo: What Nature Can Teach Us about Living Longer, Healthier Lives* (Cambridge, MA: MIT Press, 2022), 258–59.

25 **Moreover, scientists have found:** R. K. Mortimer and J. R. Johnston, "Life Span of Individual Yeast Cells," *Nature* 183, no. 4677 (June 20, 1959): 1751–52, https://doi.org/10.1038/1831751a0; E. J. Stewart et al., "Aging and Death in an Organism That Reproduces by Morphologically Symmetric Division." *PLoS Biology* 3, no. 2 (February 2005): e45, https://doi.org/10.1371/journal.pbio.0030045.

2. Live Fast and Die Young

27 **A small aquatic animal:** T. C. Bosch, "Why Polyps Regenerate and We Don't: Towards a Cellular and Molecular Framework for *Hydra* Regeneration," *Developmental Biology* 303, no. 2 (March 15, 2007): 421–33, https://doi.org/10.1016/j.ydbio.2006.12.012.

27 **Still, it is a complex procedure:** R. Murad et al., "Coordinated Gene Expression and Chromatin Regulation During *Hydra* Head Regeneration," *Genome Biology and Evolution* 13, no. 12 (December 2021): evab221, https://doi.org/10.1093/gbe/evab221; see also a popular account of this work and *hydra* in general in Corryn Wetzel, "How Tiny, 'Immortal' Hydras Regrow Their Lost Heads," *Smithsonian* online, last modified December 13, 2021, https://www.smithsonian mag.com/smart-news/were-closer-to-understanding-how-immortal -hydras-regrow-lost-heads-180979209/.

27 **It is almost as if an injured butterfly:** Y. Matsumoto and M. P. Miglietta, "Cellular Reprogramming and Immortality: Expression Profiling Reveals Putative Genes Involved in *Turritopsis dohrnii*'s Life Cycle Reversal," *Genome Biology and Evolution* 13, no. 7 (July 2021): evab136, https://doi.org/10.1093/gbe/evab136; M. Pascual-Torner et al., "Comparative Genomics of Mortal and Immortal Cnidarians

Unveils Novel Keys Behind Rejuvenation," *Proceedings of the National Academy of Sciences (PNAS) of the United States of America* 119, no. 36 (September 6, 2022): e2118763119, https://doi.org/10.1073/pnas.2118763119; see also a popular account by Veronique Greenwood, "This Jellyfish Can Live Forever. Its Genes May Tell Us How," *New York Times* online, September 6, 2022, https://www.nytimes.com/2022/09/06/science/immortal-jellyfish-gene-protein.html.

28   **Along the way, he explores:** West, *Scale*. Many of the original findings for relationships between longevity, size, and metabolic rates can be found here.

31   **As a result, biologists do not think:** For a biologist's view of the second law of thermodynamics and the wear-and-tear theory of aging, see Tom Kirkwood, chap. 5, "The Unnecessary Nature of Ageing," in *Time of Our Lives: The Science of Human Aging* (New York: Oxford University Press, 1999), 52–62.

31   **From there, he became interested:** See Austad's academic website: University of Alabama at Birmingham online, College of Arts and Science, Department of Biology, https://www.uab.edu/cas/biology/people/faculty/steven-n-austad; see also a description about him and a podcast interview, https://blog.insidetracker.com/longevity-by-design-steven-austad.

32   **The LQ is the ratio:** S. N. Austad and K. E. Fischer, "Mammalian Aging, Metabolism, and Ecology: Evidence from the Bats and Marsupials," *Journal of Gerontology* 46, no. 2 (March 1991): B47–B53, https://doi.org/10.1093/geronj/46.2.b47.

33   **Over the years, Austad has studied:** Austad, *Methuselah's Zoo*. There is also a previous short and more technical version of this: S. N. Austad, "Methusaleh's Zoo: How Nature Provides Us with Clues for Extending Human Health Span," *Journal of Comparative Pathology* 142, suppl. 1 (January 2010): S10–S21, https://doi.org/10.1016/j.jcpa.2009.10.024. Much of this section on the life span of various animals is from these two sources.

34   **Two studies that evaluated survival data:** B. A. Reinke et al., "Diverse Aging Rates in Ectothermic Tetrapods Provide Insights for the Evolution of Aging and Longevity," *Science* 376, no. 6600 (June 23, 2022): 1459–66, https://doi.org/10.1126/science.abm0151; R. da Silva et al., "Slow and Negligible Senescence Among Testudines Challenges Evolutionary Theories of Senescence," *Science* 376, no. 6600 (June 23, 2022): 1466–70, https://doi.org/10.1126/science.abl7811.

34 **By the time a person:** "Actuarial Life Table," Social Security Administration online, accessed August 7, 2023, https://www.ssa.gov/oact/STATS/table4c6.html.

34 **Like elderly humans:** S. N. Austad and C. E. Finch, "How Ubiquitous Is Aging in Vertebrates?," *Science* 376, no. 6600 (June 23, 2022): 1384–85, https://doi.org/10.1126/science.adc9442; Finch is quoted in Jack Tamisiea, "Centenarian Tortoises May Set the Standard for Anti-aging," *New York Times* online, June 23, 2022, https://www.nytimes.com/2022/06/23/science/tortoises-turtles-aging.html.

35 **Bats do not live as long:** G. S. Wilkinson and J. M. South, "Life History, Ecology and Longevity in Bats," *Aging Cell* 1, no. 2 (December 2002): 124–31, https://doi.org/10.1046/j.1474-9728.2002.00020.x.

36 **Austad estimates that its LQ:** A. J. Podlutsky et al., "A New Field Record for Bat Longevity," *Journals of Gerontology: Series A* 60, no. 11 (November 2005): 1366–68, https://doi.org/10.1093/gerona/60.11.1366.

36 **But even bats that don't hibernate:** Wilkinson and South, "Life History," 124–31.

36 **Rather, they may have special mechanisms:** Podlutsky et al., "New Field Record," 1366–68.

37 **Rochelle Buffenstein, currently at the University of Illinois in Chicago, has done more:** R. Buffenstein, "The Naked Mole-Rat: A New Long-Living Model for Human Aging Research," *Journals of Gerontology: Series A* 60, no. 11 (November 2005): 1366–77, https://doi.org/10.1093/gerona/60.11.1369.

37 **Instead of proliferating:** S. Liang et al., "Resistance to Experimental Tumorigenesis in Cells of a Long-Lived Mammal, the Naked Mole-Rat (*Heterocephalus glaber*)," *Aging Cell* 9, no. 4 (August 2010): 626–35, https://doi.org/10.1111/j.1474-9726.2010.00588.x.

37 **One of the biggest headlines:** J. G. Ruby, M. Smith, and R. Buffenstein, "Naked Mole-Rat Mortality Rates Defy Gompertzian Laws by Not Increasing with Age," *eLife* 7 (January 24, 2018): e31157, https://doi.org/10.7554/eLife.31157.

37 **This was too much for some scientists:** S. Braude et al., "Surprisingly Long Survival of Premature Conclusions About Naked Mole-Rat Biology," *Biological Reviews of the Cambridge Philosophical Society* 96, no. 2 (April 2021): 376–93, https://doi.org/10.1111/brv.12660.

37 **As we saw with long-lived tortoises:** R. Buffenstein, et al., "The Naked Truth: A Comprehensive Clarification and Classification of Current 'Myths' in Naked Mole-Rat Biology," *Biological Reviews of* *the Cambridge Philosophical Society* 97, no. 1 (February 2022): 115–40, https://doi.org/10.1111/brv.12791.

38 **The science writer Steven Johnson:** Steven Johnson, *Extra Life: A Short History of Living Longer* (New York: Riverhead Books, 2021).

39 **The ability to chemically capture nitrogen:** The dramatic impact of fertilizers on humanity is told in Thomas Hager's fascinating book *The Alchemy of Air: A Jewish Genius, a Doomed Tycoon, and the Scientific Discovery That Fed the World but Fueled the Rise of Hitler* (New York: Crown, 2009).

40 **He and his colleagues contended:** S. J. Olshansky, B. A. Carnes, and C. Cassel. "In Search of Methuselah: Estimating the Upper Limits to Human Longevity," *Science* 250, no. 4981 (November 2, 1990): 634–40, https://doi.org/10.1126/science.2237414; S. J. Olshansky, B. A. Carnes, and A. Désesquelles, "Prospects for Human Longevity," *Science* 291, no. 5508 (February 23, 2001): 1491–92, https://doi.org/10.1126/science.291.5508.1491.

40 **Moreover, in certain species:** A. Baudisch and J. W. Vaupel, "Getting to the Root of Aging: Why Do Patterns of Aging Differ Widely Across the Tree of Life?," *Science* 338, no. 6107 (November 2, 2012): 618–19, https://doi.org/10.1126/science.1226467; O. R. Jones and J. W. Vaupel, "Senescence Is Not Inevitable," *Biogerontology* 18, no. 6 (December 2017): 965–71, https://doi.org/10.1007/s10522-017 -9727-3.

40 **The disagreements between the two boiled:** See J. Couzin-Frankel, "A Pitched Battle over Life Span," *Science* 338, no. 6042 (July 29, 2011): 549–50, https://doi.org/10.1126/science.333.6042.549.

40 **"pernicious belief":** J. Oeppen and J. W. Vaupel, "Demography. Broken Limits to Life Expectancy," *Science* 296, no. 5570 (May 10, 2022): 1029–1031, https://doi.org/10.1126/science.1069675.

40 **In agreement with this:** F. Colchero et al., "The Long Lives of Primates and the 'Invariant Rate of Ageing' Hypothesis," *Nature Communications* 12, no. 1 (June 16, 2021): 3666, https://doi.org/10.1038 /s41467-021-23894-3.

41 **Unlike most people:** There is an entertaining account of Parr in Austad, *Methuselah's Zoo*, pages 262–63.

41 **"Until next year, perhaps":** Craig R. Whitney, "Jeanne Calment, World's Elder, Dies at 122," *New York Times*, August 5, 1997, B8.

42 **Vijg predicted:** X. Dong, B. Milholland, and J. Vijg, "Evidence for a Limit to Human Lifespan," *Nature* 538, no. 7624 (October 13, 2016): 257–59, https://doi.org/10.1038/nature19793.

42 **"if any":** E. Barbi et al., "The Plateau of Human Mortality: Demography of Longevity Pioneers," *Science* 360, no. 6396 (June 29, 2018): 1459–61, https://doi.org/10.1126/science.aat3119.

42 **This paper in turn was criticized:** Carl Zimmer, "How Long Can We Live? The Limit Hasn't Been Reached, Study Finds," *New York Times* online, June 28, 2018, https://www.nytimes.com/2018/06/28/science/human-age-limit.html.

42 **Others pointed out:** H. Beltrán-Sánchez, S. N. Austad, and C. E. Finch, "The Plateau of Human Mortality: Demography of Longevity Pioneers," *Science* 361, no. 6409 (September 28, 2018): eaav1200, https://doi.org/10.1126/science.aav1200.

43 **After climbing steadily for the last 150 years:** C. Cardona and D. Bishai, "The Slowing Pace of Life Expectancy Gains Since 1950," *BMC Public Health* 18, no. 1 (January 17, 2018): 151, https://doi.org/10.1186/s12889-018-5058-9; J. Schöley et al., "Life Expectancy Changes Since COVID-19," *Nature Human Behaviour* 6, no. 12 (December 2022): 1649–59, https://doi.org/10.1038/s41562-022-01450-3.

43 **As I write this:** "List of the Verified Oldest People," Wikipedia, last accessed July 10, 2023, https://en.wikipedia.org/wiki/List_of_the_verified_oldest_people.

44 **In fact, about half of centenarians:** J. Evert et al., "Morbidity Profiles of Centenarians: Survivors, Delayers, and Escapers," *Journals of Gerontology: Series A, Biological Sciences and Medical Sciences* 58, no. 3 (March 2003): 232–37, https://doi.org/10.1093/gerona/58.3.m232.

44 **He agrees with Olshansky:** Thomas Perls, email messages to the author, November 27, 2021, and January 17, 2022.

45 **A dozen years later:** Described in Austad, *Methuselah's Zoo*, 273–74.

46 **But scientists have homed in:** C. López-Otín et al., "The Hallmarks of Aging," *Cell* 153, no. 6 (June 6, 2013): 1194–217, https://doi.org/10.1016/j.cell.2013.05.039. This classic paper has recently been updated on the tenth anniversary of the original: C. López-Otín et al. "Hallmarks of Aging: An Expanding Universe," *Cell* 186, no. 1 (January 19, 2023): 243–78, https://doi.org/10.1016/j.cell.2022.11.001.

3. Destroying the Master Controller

49 **Today we know that our genes:** Two very readable accounts of the history of genetics can be found in Matthew Cobb, *Life's Greatest Secret: The Race to Crack the Genetic Code* (London: Profile Books, 2015), and Siddhartha Mukherjee, *The Gene: An Intimate History* (New York: Scribner, 2017).

51   **How instructions in mRNA are read:** The decade-long effort to crack the genetic code and understand how proteins are made is described in Cobb, *Life's Greatest Secret*.

52   **I have spent much of my life:** Venki Ramakrishnan, *Gene Machine: The Race to Decipher the Secrets of the Ribosome* (London: Oneworld, 2018).

55   **As early as the eighteenth century:** H. W. Herr, "Percivall Pott, the Environment and Cancer," *BJU International* 108, no. 4 (August 2011): 479–81, https://doi.org/10.1111/j.1464-410x.2011.10487.x.

55   **Hermann Muller was a third-generation American who grew up in New York City:** G. Pontecorvo, "Hermann Joseph Muller, 1890–1967," *Biographical Memoirs of Fellows of the Royal Society* 14 (November 1968): 348–89, https://doi.org/10.1098/rsbm.1968.0015; Elof Axel Carlson, *Hermann Joseph Muller 1890–1967: A Biographical Memoir* (Washington, DC: National Academy of Sciences, 2009), available at http://www.nasonline.org/publications/biographical-memoirs/memoir-pdfs/muller-hermann.pdf.

55   **Even a modest application:** Errol Friedberg, chap. 1, "In the Beginning," in *Correcting the Blueprint of Life: An Historical Account of the Discovery of DNA Repair Mechanisms* (Cold Spring Harbor, NY: Cold Spring Harbor Laboratory Press, 1997).

57   **One of Crew's key collaborators:** Geoffrey Beale, "Charlotte Auerbach, 14 May 1899–1917 March 1994," *Biographical Memoirs of Fellows of the Royal Society* 41 (November 1995): 20–42, https://doi.org/10.1098/rsbm.1995.0002

58   **But once Watson and Crick revealed its double-helical nature:** A very good historical summary of early work on DNA damage and repair can be found in Friedberg, chap. 1, "In the Beginning," in *Correcting the Blueprint of Life*.

60   **Sunlight could kill bacteria:** A. Downes and T. P. Blunt, "The Influence of Light upon the Development of Bacteria," *Nature*, 16 (July 12, 1877), 218, https://doi.org/10.1038/016218a0; F. L. Gates, "A Study of the Bactericidal Action of Ultraviolet Light," *Journal of General Physiology*, 14, No. 1 (September 20, 1930): 31–42, https://doi.org/10.1085/jgp.14.1.31.

60   **However, when they tried this:** R. B. Setlow and J. K. Setlow, "Evidence That Ultraviolet-Induced Thymine Dimers in DNA Cause Biological Damage," *Proceedings of the National Academy of Sciences*

(PNAS) of the United States of America 48, no. 7 (July 1, 1962): 1250–57, https://doi.org/10.1073/pnas.48.7.1250.

60   **Dick and his colleagues found:** R. B. Setlow, P. A. Swenson, and W. L. Carrier, "Thymine Dimers and Inhibition of DNA Synthesis by Ultraviolet Irradiation of Cells," *Science* 142, no. 3698 (December 13, 1963): 1464–66, https://doi.org/10.1126/science.142.3598.1464; R. B. Setlow and W. L. Carrier, "The Disappearance of Thymine Dimers from DNA: An Error-Correcting Mechanism, *Proceedings of the National Academy of Sciences (PNAS) of the United States of America* 51, no. 2 (April 1964): 226–31, https://doi.org/10.1073/pnas.51.2.226.

61   **The same year:** R. P. Boyce and P. Howard-Flanders, "Release of Ultraviolet Light-Induced Thymine Dimers from DNA in *E. coli* K-12," *Proceedings of the National Academy of Sciences (PNAS) of the United States of America* 51, no. 2 (February 1, 1964): 293–300, https://doi.org/10.1073/pnas.51.2.293; D. Pettijohn and P. Hanawalt, "Evidence for Repair-Replication of Ultraviolet Damaged DNA in Bacteria," *Journal of Molecular Biology* 9, no. 2 (August 1964): 395–410, https://doi.org/10.1016/s0022-2836(64)80216-3.

61   **How it worked was something of a mystery:** Aziz Sancar, "Mechanisms of DNA Repair by Photolyase and Excision Nuclease (Nobel Lecture, December 8, 2015), available at https://www.nobelprize.org/uploads/2018/06/sancar-lecture.pdf.

62   **That is a *very* long time:** A great account of Thomas Lindahl's discoveries can be found in his "The Intrinsic Fragility of DNA" (Nobel Lecture, December 8, 2015), available at https://www.nobelprize.org/uploads/2018/06/lindahl-lecture.pdf.

63   **Lindahl estimated later:** Tomas Lindahl, "Instability and Decay of the Primary Structure of DNA," *Nature* 362, no. 6422 (April 22, 1993): 709–715.

64   **Not surprisingly, the cell:** Paul Modrich, "Mechanisms in *E. coli* and Human Mismatch Repair" (Nobel Lecture, December 8, 2015, https://www.nobelprize.org/uploads/2018/06/modrich-lecture.pdf).

64   **Relying on some very clever experiments:** Ibid.

65   **The prize also cannot be given:** As is increasingly the case because of the limitation of the Nobel Prize to three people, the prize for DNA repair was not without its controversy: David Kroll, "This Year's Nobel Prize in Chemistry Sparks Questions About How Winners Are Selected," *Chemical & Engineering News (C&EN)* online, last modified November 11, 2015, https://cen.acs.org/articles/93/i45/Years-Nobel-Prize-Chemistry-Sparks.html.

65 **One condition he has focused on:** B. Schumacher et al., "The Central Role of DNA Damage in the Ageing Process," Nature 592, no. 7856 (April 2021): 695–703, https://doi.org/10.1038/s41586 -021-03307-7.

66 **In females, defects in how the cell:** K. T. Zondervan, "Genomic Analysis Identifies Variants That Can Predict the Timing of Menopause," *Nature* 596, no. 7872 (August 2021): 345–46, https://doi .org/10.1038/d41586-021-01710-8; K. S. Ruth et al., "Genetic Insights into Biological Mechanisms Governing Human Ovarian Ageing," *Nature* 596, no. 7872 (August 2021): 393–97, https://doi .org/10.1038/s41586-021-03779-7. See also the commentary by H. Ledford, "Genetic Variations Could One Day Help Predict Timing of Menopause," *Nature* online, last modified August 4, 2021, https://doi.org/10.1038/d41586-021-02128-y.

66 **Sometimes the cell:** Apoptosis, or programmed cell death, is also a feature of normal development, as specific cells die at precise points during the development of an organism from a single cell into the adult animal. This was first discovered by studying how the worm *C. elegans* develops from a single fertilized egg into an adult of almost a thousand cells, and resulted in the award of the 2002 Nobel Prize to Sydney Brenner, John Sulston, and Robert Horvitz.

67 **When the damage is too extensive:** A. J. Levine and G. Lozano, eds., *The P53 Protein: From Cell Regulation to Cancer*, Cold Spring Harbor Perspectives in Medicine (Cold Spring Harbor, NY: Cold Spring Harbor Laboratory, 2016).

67 **Humans inherit one copy:** L. M. Abegglen et al., "Potential Mechanisms for Cancer Resistance in Elephants and Comparative Cellular Response to DNA Damage in Humans," *Journal of the American Medical Association* (*JAMA*) 314, no. 17 (November 3, 2015): 1850–60, https://doi.org/10.1001/jama.2015.13134; M. Sulak et al., "TP53 Copy Number Expansion Is Associated with the Evolution of Increased Body Size and an Enhanced TP Damage Response in Elephants," *eLife* 5 (2016): e11994, https://doi.org/10.7554/eLife.11994.

67 **Curiously, in studies:** M. Shaposhnikov et al., "Lifespan and Stress Resistance in *Drosophila* with Overexpressed DNA Repair Genes," *Scientific Reports* 5 (October 19, 2015): art. 15299, https://doi.org /10.1038/srep15299.

67 **Some of the long-lived species:** D. Tejada-Martinez, J. P. de Magalhães, and J. C. Opazo, "Positive Selection and Gene Duplications in Tumour Suppressor Genes Reveal Clues About How

Cetaceans Resist Cancer," *Proceedings of the Royal Society B* (*Biological Sciences*) 288, no. 1945 (February 24, 2021): art. 20202592, https://doi.org/10.1098/rspb.2020.2592; V. Quesada et al., "Giant Tortoise Genomes Provide Insights into Longevity and Age-Related Disease," *Nature Ecology & Evolution* 3 (January 2019): 87–95, https://doi.org/10.1038/s41559-018-0733-x.

67   **Humans and naked mole rats:** S. L. MacRae et al., "DNA Repair in Species with Extreme Lifespan Differences," *Aging* 7, no. 12 (December 2015): 1171–84, https://doi.org/10.18632/aging.100866.

68   **Paradoxically, many new cancer therapies:** See, for example, Liam Drew, "PARP Inhibitors: Halting Cancer by Halting DNA Repair," Cancer Research UK online, last modified September 24, 2020, https://news.cancerresearchuk.org/2020/09/24/parp-inhibitors-halting-cancer-by-halting-dna-repair/.

4. The Problem with Ends

70   **"Perhaps the day":** *Scientific American*, July 1921, quoted in Mark Fischetti, comp., "1921: Immortality for Humans," *Scientific American* online, July 2021, 79, https://robinsonlab.cellbio.jhmi.edu/wp-content/uploads/2021/06/SciAm_2021_07.pdf.

70   **They were not immortal:** An engaging history of Hayflick's discovery and its aftermath is J. W. Shay and W. E. Wright, "Hayflick, His Limit, and Cellular Ageing," *Nature Reviews Molecular Cell Biology* 1, no. 1 (October 2000): 72–76, https://doi.org/10.1038/35036093.

71   **It has since become a classic:** L. Hayflick and P. S. Moorhead, "The Serial Cultivation of Human Diploid Cell Strains," *Experimental Cell Research* 25, no. 3 (December 1961): 585–621, https://doi.org/10.1016/0014-4827(61)90192-6.

71   **Some have even suggested:** J. Witkowski, "The Myth of Cell Immortality," *Trends in Biochemical Sciences* 10, no. 7 (July 1985): 258–60, https://doi.org/10.1016/0968-0004(85)90076-3.

72   **Given Carrel's stature:** John J. Conley, "The Strange Case of Alexis Carrel, Eugenicist," in *Life and Learning XXIII and XXIV: Proceedings of the Twenty-third (2013) and Twenty-fourth Conferences of the University Faculty for Life Conference at Marquette University, Milwaukee, Wisconsin*, vol. 26, ed. Joseph W. Koterski (Milwaukee: University Faculty for Life), 281–88, https://www.uffl.org/pdfs/vol23/UFL_2013_Conley.pdf.

72   **Titia de Lange:** Titia de Lange, conversation with the author, September 10, 2021.

72   **He realized that the train:** This so-called end replication problem was first pointed out by J. D. Watson, "Origin of Concatemeric T7 DNA," *Nature New Biology* 239, no. 94 (October 18, 1972): 197–201, https://doi.org/10.1038/newbio239197a0, and A. M. Olovnikov, "Telomeres, Telomerase, and Aging: Origin of the Theory," *Experimental Gerontology* 31, no. 4 (July/August 1996): 443–48, https://www.sciencedirect.com/science/article/abs/pii/0531556596000058. For a good description of how it would work, see M. M. Cox, J. Doudna, and M. O'Donnell, *Molecular Biology: Principles and Practice* (New York: W. H. Freeman, 2012), 398–400. The Wikipedia page "DNA Replication," last modified June 14, 2023, https://en.wikipedia.org/wiki/DNA_replication, is also quite informative.

73   **At some point, she discovered:** For a long time, McClintock was not believed, but these so-called transposable elements turned out to be a fundamental part of biology, and she was awarded the Nobel Prize for her work in 1983 at the age of eighty-one.

74   **TTGGGG:** E. H. Blackburn and J. G. Gall, "A Tandemly Repeated Sequence at the Termini of the Extrachromosomal Ribosomal RNA Genes in Tetrahymena," *Journal of Molecular Biology* 120, no. 1 (March 25, 1978): 33–53, https://doi.org/10.1016/0022-2836(78)90294-2.

75   **It worked like a charm:** J. W. Szostak and E. H. Blackburn, "Cloning Yeast Telomeres on Linear Plasmid Vectors," *Cell* 29, no. 1 (May 1982): 245–55, https://doi.org/10.1016/0092-8674(82)90109-x.

76   **The two of them discovered an enzyme:** C. W. Greider and E. H. Blackburn, "Identification of a Specific Telomere Terminal Transferase Activity in Tetrahymena Extracts," *Cell* 43, no. 2, pt. 1 (November 1985): 405–13, https://doi.org/10.1016/0092-8674(85)90170-9; C. W. Greider and E. H. Blackburn, "The Telomere Terminal Transferase of Tetrahymena Is a Ribonucleoprotein Enzyme with Two Kinds of Primer Specificity," *Cell* 51, no. 6 (December 24, 1987): 887–98, https://doi.org/10.1016/0092-8674(87)90576-9; C. W. Greider and E. H. Blackburn, "A Telomeric Sequence in the RNA of Tetrahymena Telomerase Required for Telomere Repeat Synthesis," *Nature* 337, no. 6205 (January 26, 1989): 331–37, https://doi.org/10.1038/337331a0.

76   **Without telomerase:** C. B. Harley, A. B. Futcher, and C. W. Greider, "Telomeres Shorten During Ageing of Human Fibroblasts," *Nature* 345, no. 5274 (May 31, 1990): 458–60, https://doi.org/10.1038/345458a0.

76   **Even introducing telomerase:** A. G. Bodnar et al., "Extension of Life-span by Introduction of Telomerase into Normal Human

Cells," *Science* 279, no. 5349 (January 16, 1998): 349–52, https://doi .org/10.1126/science.279.5349.349.

76 **It turns out that the telomeric ends:** The strand that extends beyond the other is called a 3' overhang, so the reason for the loss of the ends is not exactly the reason first proposed by Olovnikov and Watson. Aficionados can look at J. Lingner, J. P. Cooper, and T. R. Cech, "Telomerase and DNA End Replication: No Longer a Lagging Strand Problem," *Science* 269, no. 5230 (September 15, 1995): 1533–34, https://doi.org/10.1126/science.7545310.

76 **This longer strand:** T. de Lange, "Shelterin: The Protein Complex That Shapes and Safeguards Human Telomeres," *Genes & Development* 19, no. 18 (September 15, 2005): 2100–10, https://doi.org/10.1101/ gad.1346005; I. Schmutz and T. de Lange, "Shelterin," *Current Biology* 26, no. 10 (May 23, 2016): R397–99, https://doi.org/10.1016 /j.cub.2016.01.056.

76 **This crucial structure is why the cell:** W. Palm and T. de Lange, "How Shelterin Protects Mammalian Telomeres," *Annual Review of Genetics* 42 (2008): 301–34, https://doi.org/10.1146/annurev. genet.41.110306.130350; P. Martínez and M. A. Blasco, "Role of Shelterin in Cancer and Aging," *Aging Cell* 9, no. 5 (October 2010): 653–66, https://doi.org/10.1111/j.1474-9726.2010.00596.x.

76 **The cell then sees:** F. d'Adda di Fagagna et al. "A DNA Damage Checkpoint Response in Telomere-Initiated Senescence," *Nature* 426, no. 6963 (November 13, 2003): 194–98, https://doi.org/10.1038 /nature02118.

77 **People with defective telomerase:** M. Armanios and E. H. Blackburn, "The Telomere Syndromes," *Nature Reviews Genetics* 13, no. 10 (October 2012): 693–704, https://doi.org/10.1038/nrg3246.

77 **When we are stressed:** E. S. Epel et al., "Accelerated Telomere Shortening in Response to Life Stress," *Proceedings of the National Academy of Sciences (PNAS) of the United States of America* 101, no. 49 (December 1, 2004): 17312–15, https://doi.org/10.1073/pnas.0407162101; J. Choi, S. R. Fauce, and R. B. Effros, "Reduced Telomerase Activity in Human T Lymphocytes Exposed to Cortisol," *Brain, Behavior, and Immunity* 22, no. 4 (May 2008): 600–605, https://doi.org/10.1016 /j.bbi.2007.12.004. See also the following on stress and premature gray hair in mice: B. Zhang et al., "Hyperactivation of Sympathetic Nerves Drives Depletion of Melanocyte Stem Cells," *Nature* 577, no. 792 (January 2020): 676–81, https://doi.org/10.1038/s41586-020 -1935-3.

77   **So it may be that the shortening:** M. Jaskelioff et al. "Telomerase
      Reactivation Reverses Tissue Degeneration in Aged Telomerase-
      Deficient Mice," *Nature* 469, no. 7328 (January 6, 2001): 102–6
      (2011), https://doi.org/10.1038/nature09603.

77   **According to a number of studies, mice engineered:** M. A. Muñoz-
      Lorente, A. C. Cano-Martin, and M. A. Blasco, "Mice with Hyper-
      long Telomeres Show Less Metabolic Aging and Longer Lifespans,"
      Nature Communications 10, no. 1 (October 17, 2019): 4723, https://doi
      .org/10.1038/s41467-019-12664-x.

78   **There seems to be a delicate balance:** Titia de Lange, conversa-
      tions with and email messages to the author, November and Decem-
      ber 2021. See also Jalees Rehman, "Aging: Too Much Telomerase
      Can Be as Bad as Too Little," Guest Blog, *Scientific American* online,
      last modified July 5, 2014, ttps://blogs.scientificamerican.com/guest
      -blog/aging-too-much-telomerase-can-be-as-bad-as-too-little/.

78   **On the other hand, those with long telomeres:** E. J. McNally, P. J.
      Luncsford, and M. Armanios, "Long Telomeres and Cancer Risk: The
      Price of Cellular Immortality," *Journal of Clinical Investigation* 129,
      no. 9 (August 5, 2019): 3474–81, https://doi.org/10.1172/JCI120851.

5. Resetting the Biological Clock

79   **"another great Anglo-American partnership":** The official text
      of the statement on the publication of the draft human genome
      sequence by the White House and the UK government is here: Na-
      tional Human Genome Research Institute online, "June 2000 White
      House Event," news release, June 26, 2000, https://www.genome.
      gov/10001356/june-2000-white-house-event. A slightly different text
      was reported by the *New York Times*: "Text of the White House State-
      ments on the Human Genome Project," Science, *New York Times* on-
      line, June 27, 2000, https://archive.nytimes.com/www.nytimes.com/
      library/national/science/062700sci-genome-text.html. The sequence
      itself was described in two large, coordinated publications: the public
      consortium was published as International Human Genome Se-
      quencing Consortium et al., "Initial Sequencing and Analysis of
      the Human Genome," *Nature* 409, no. 6822 (February 15, 2001):
      860–921, https://doi.org/10.1038/35057062, while the private Celera
      effort was published as J. C. Venter et al., "The Sequence of the
      Human Genome," *Science* 291, 1304–51, https://doi.org/10.1126
      /science.1058040.

79 **"Along with Bach's music":** Quoted in G. Yamey, "Scientists Unveil First Draft of Human Genome," *BMJ* 321, no. 7252 (July 1, 2000): 7, https://doi.org/10.1136/bmj.321.7252.7.

80 **Venter was something:** "Profile: Craig Venter," BBC News online, last modified May 21, 2010, https://www.bbc.co.uk/news/10138849.

80 **The decision by NIH:** "US Patent Application Stirs Up Gene Hunters," *Nature*, 353 (October 10, 1991): 485–86 (1991), https://doi.org/10.1038/353485a0; N. D. Zinder, "Patenting cDNA 1993: Efforts and Happenings" (abstract), *Gene* 135, nos. 1/2 (December 1993): 295–98, https://www.sciencedirect.com/science/article/abs/pii/037811199390080M.

80 **Venter said later that he was always against them:** Matthew Herper, "Craig Venter Mapped the Genome. Now He's Trying to Decode Death," *Forbes* (online), February 21, 2017, https://www.forbes.com/sites/matthewherper/2017/02/21/can-craig-venter-cheat-death/?sh=8f6fefa16456.

81 **A particularly passionate advocate:** John Sulston and Georgina Ferry, *The Common Thread: A Story of Science, Politics, Ethics, and the Human Genome* (New York: Random House, 2002).

81 **In the run-up:** "How Diplomacy Helped to End the Race to Sequence the Human Genome," *Nature* 582, no. 7813 (June 2020): 460, https://doi.org/10.1038/d41586-020-01849-w.

81 **The sequence was declared finished:** S. Reardon, "A Complete Human Genome Sequence Is Close: How Scientists Filled in the Gaps," *Nature* 594, no. 7862 (June 2021): 158–59, https://doi.org/10.1038/d41586-021-01506-w.

83 **The study of this change:** Nessa Carey's *The Epigenetics Revolution: How Modern Biology Is Rewriting Our Understanding of Genetics, Disease, and Inheritance* (New York: Columbia University Press, 2012) is a great popular introduction to epigenetics. Mukherjee's *The Gene* is more broadly about the nature of the gene but has a significant emphasis on epigenetics.

86 **They are too far down:** R. Briggs and T. J. King, "Transplantation of Living Nuclei from Blastula Cells into Enucleated Frogs' Eggs," *Proceedings of the National Academy of Sciences (PNAS) of the United States of America* 38, no. 5 (May 1952): 455–63, https://doi.org/10.1073/pnas.38.5.455.

87 **He studied languages instead:** "Sir John B. Gurdon: Biographical," Nobel Prize online, accessed August 7, 2023, https://www.nobelprize.org/prizes/medicine/2012/gurdon/biographical/.

88   **The clawed frog became:** J. B. Gurdon and N. Hopwood, "The Introduction of Xenopus Laevis into Developmental Biology: Of Empire, Pregnancy Testing and Ribosomal Genes," *International Journal of Developmental Biology* 44, no. 1 (2000): 43–50.

88   **This was the first time:** J. B. Gurdon, "The Developmental Capacity of Nuclei Taken from Intestinal Epithelium Cells of Feeding Tadpoles," *Development* 10, no. 4 (December 1, 1962): 622–40, https://doi .org/10.1242/dev.10.4.622.

89   **Eventually other researchers reproduced:** I. Wilmut et al., "Viable Offspring Derived from Fetal and Adult Mammalian Cells," *Nature* 385, no. 6619 (February 27, 1997): 810–13, https://doi.org /10.1038/385810a0.

90   **Being able to grow ES cells:** M. J. Evans and M. H. Kaufman, "Establishment in Culture of Pluripotential Cells from Mouse Embryos," *Nature* 292, no. 5819 (July 9, 1981): 154–56, https://doi .org/10.1038/292154a0; G. R. Martin, "Isolation of a Pluripotent Cell Line from Early Mouse Embryos Cultured in Medium Conditioned by Teratocarcinoma Stem Cells," *Proceedings of the National Academy of Sciences (PNAS) of the United States of America* 78, no. 12 (December 1, 1981): 7634–38, https://doi.org/10.1073/pnas.78.12.7634.

92   **By experimenting with transcription factors in various combinations:** Shinya Yamanaka, "Shinya Yamanaka: Biographical," Nobel Prize online, https://www.nobelprize.org/prizes/medicine/2012 /yamanaka/biographical/.

93   **One of the first and simplest:** The *lac* operator and repressor system was discovered in the 1960s by Jacques Monod and Francois Jacob, and its history, along with another genetic switch in a bacteriophage by Andre Lwoff, resulted in the Nobel Prize in 1965. For an insightful history, see M. Lewis, "A Tale of Two Repressors," *Journal of Molecular Biology* 409, no. 1 (May 27, 2011): 14–27, https://doi.org/10.1016 /j.jmb.2011.02.023.

94   **You might expect that when cells divide:** The British geneticist Adrian Bird showed that the methylation occurs mainly on islands with CG repeats. Because C pairs with a G, if you have a CpG island, the C and G on each strand will be directly across from a G and C on the opposite strand. Each C will then be diagonally across from the C on the other strand. When cells methylate a CpG island, they methylate the Cs on both strands. As soon as the cell divides, you have two molecules of DNA instead of one. Each of them has an original strand where the C is methylated, and a newly made strand in which it isn't. There are special methyltransferase enzymes that will add a methyl group to a C only if the C diagonally across from it on the other strand already has one. This ensures that both strands end up methylated exactly in the same places they were before.

95 **It is a striking example:** E. W. Tobi et al., "DNA Methylation as a Mediator of the Association Between Prenatal Adversity and Risk Factors for Metabolic Disease in Adulthood," *Science Advances* 4, no. 1 (January 31, 2018): eaao4364, https://doi.org/10.1126/sciadv .aao4364; described in Carl Zimmer, "The Famine Ended 70 Years Ago, But Dutch Genes Still Bear Scars," *New York Times* online, January 31, 2018, https://www.nytimes.com/2018/01/31/science/dutch -famine-genes.html. See also Mukherjee, The Gene, and Carey, *The Epigenetics Revolution.*

97 **When they looked at the methylation:** For an expert popular account of Steve Horvath and epigenetic clocks, see Ingrid Wickelgren, "Epigenetic 'Clocks' Predict Animals' True Biological Age," Quanta, last modified August 17, 2022, https://www.quantamagazine.org /epigenetic-clocks-predict-animals-true-biological-age-20220817/. Some of the background on Horvath is taken from this article.

97 **He was able to identify 513 sites:** M. E. Levine et al., "An Epigenetic Biomarker of Aging for Lifespan and Healthspan," *Aging* 10, no. 4 (April 2018): 573–91, https://doi.org/10.18632/aging.101414.

98 **Methylation patterns are like a biological clock:** S. Horvath and K. Raj, "DNA Methylation-Based Biomarkers and the Epigenetic Clock Theory of Ageing," *Nature Reviews Genetics* 19, no. 6 (June 2018): 371–84, https://doi.org/10.1038/s41576-018-0004-3.

98 **Many other research groups developed:** For an example, see G. Hannum et al., "Genome-wide Methylation Profiles Reveal Quantitative Views of Human Aging Rates," *Molecular Cell* 49, no. 2 (January 24, 2013): 359–67, https://doi.org/10.1016/j.mol cel.2012.10.016.

98 **In fact, its methylation pattern:** C. Kerepesi et al., "Epigenetic Clocks Reveal a Rejuvenation Event During Embryogenesis Followed by Aging," *Science Advances* 7, no. 26 (June 25, 2021): eabg6082, https://doi.org/10.1126/sciadv.abg6082; C. Kerepesi et al., "Epigenetic Aging of the Demographically Non-Aging Naked Mole-Rat," *Nature Communications* 13, no. 1 (January 17, 2022): 355, https://doi .org/10.1038/s41467-022-27959-9.

99 **Something about her diet:** R. Kucharski et al., "Nutritional Control of Reproductive Status in Honeybees Via DNA Methylation," *Science*

319, no. 5871 (March 28, 2008): 1827–30, https://doi.org/10.1126/science.1153069; M. Wojciechowski et al., "Phenotypically Distinct Female Castes in Honey Bees Are Defined by Alternative Chromatin States During Larval Development," *Genome Research* 28, no. 10 (October 2018): 1532–42, https://doi.org/10.1101/gr.236497.118.

100 **The first is that germ-line cells:** L. Moore et al., "The Mutational Landscape of Human Somatic and Germline Cells," *Nature* 597, no. 7876 (September 2021): 381–86, https://doi.org/10.1038/s41586-021-03822-7.

100 **By puberty, this number:** Kirkwood, *Time of Our Lives*, 167–78.

100 **And even within an embryo that is developing normally overall:** A recent example is A. Lima et al., "Cell Competition Acts as a Purifying Selection to Eliminate Cells with Mitochondrial Defects During Early Mouse Development," *Nature Metabolism* 3, no. 8 (August 2021): 1091–108, https://doi.org/10.1038/s42255-021-00422-7, but there are many ways in which the body rejects defective embryos from developing to term.

101 **This is because the pronuclei:** Azim Surani, the scientist in Cambridge who first showed that a fertilized egg needed nuclei from both paternal and maternal germ-line cells to develop normally into a new animal, first suggested the idea of random, environmentally induced, and possibly deleterious epigenetic changes in our genome, which he called "epimutations." Interview with the author, February 10, 2022.

102 **There were also the lesser-known:** Joanna Klein, "Dolly the Sheep's Fellow Clones, Enjoying Their Golden Years," *New York Times* online, July 26, 2016, https://www.nytimes.com/2016/07/27/science/dolly-the-sheep-clones.html, reports on K. D. Sinclair et al., "Healthy Ageing of Cloned Sheep," *Nature Communications* 7 (July 26, 2016): 12359, https://doi.org/10.1038/ncomms12359. An extensive analysis of cloned animals in 2017 showed no systematically lower life span or other problems, suggesting that at least some cloned animals live just as long and healthy lives as naturally conceived ones: J. P. Burgstaller and G. Brem, "Aging of Cloned Animals: A Mini-Review," *Gerontology* 63, no. 5 (August 2017): 417–25, https://doi.org/10.1159/000452444.

103 **This route to rejuvenating:** T. A. Rando and H. Y. Chang, "Aging, Rejuvenation, and Epigenetic Reprogramming: Resetting the Aging Clock," *Cell* 148, no. 1/2 (January 20, 2012): 46–57, https://doi.org/10.1016/j.cell.2012.01.003; J. M. Freije and C. López-Otín, "Reprogramming Aging and Progeria," *Current Opinion in Cell Biology* 24, no. 6 (December 2012): 757–64, https://doi.org/10.1016/j.ceb.2012.08.009.

6. Recycling the Garbage

104 Today more than fifty million people: "Dementia," World Health Organization online, last modified March 15, 2023, https://www.who .int/news-room/fact-sheets/detail/dementia.

104 In England and Wales: "Dementia Now Leading Cause of Death," BBC News online, last modified November 14, 2016, https://www .bbc.co.uk/news/health-37972141.

105 It is estimated: "One-Third of British People Born in 2015 'Will Develop Dementia,'" *Guardian* (US edition) online, last modified September 21, 2015, https://www.theguardian.com/society/2015 /sep/21/one-third-of-people-born-in-2015-will-develop-dementia.

105 Over half of those with dementia: A very engaging and moving book on Alzheimer's disease is Joseph Jebelli, *In Pursuit of Memory: The Fight Against Alzheimer's* (London: John Murray, 2017). The author grew up with a grandfather who suffered from the disease.

108 There are many ways that the folding process: R. J. Ellis, "Assembly Chaperones: A Perspective," *Philosophical Transactions of the Royal Society of London, Series B, Biological Sciences* 368, no. 1617 (March 25, 2013): 20110398, https://doi.org/10.1098/rstb.2011.0398.

109 But as we age: M. Fournet, F. Bonté, and A. Desmoulière, "Glycation Damage: A Possible Hub for Major Pathophysiological Disorders and Aging," *Aging and Disease* 9, no. 5 (October 2018): 880–900, https:// doi.org/10.14336/AD.2017.1121.

109 Cells have an elaborate sensor: For an accessible description of the unfolded protein response, see Evelyn Strauss, "Unfolded Protein Response: 2014 Albert Lasker Basic Medical Research Award," Lasker Foundation online, accessed July 7, 2023, https://laskerfoundation.org /winners/unfolded-protein-response/#achievement. How exactly the sensor detects that there are too many unfolded proteins is still not entirely clear. I spoke with Dr. David Ron, a scientist at England's Cambridge Institute for Medical Research, and one of the leaders in this area. One idea is that some chaperones—the proteins that help proteins to fold—are normally abundant and can bind to the sensors, which are then kept in a quiescent state. When the number of unfolded proteins increases, these chaperones are called to action, and they release the sensors, which then go on to trigger the unfolded protein response. S. Preissler and D. Ron, "Early Events in the Endoplasmic Reticulum Unfolded Protein Response," *Cold Spring Harbor Perspectives in Biology 11*, no. 4 (April 1, 2019): a033894, https://doi.org/10.1101/cshperspect.a033894.

110 **In extreme cases:** A. Fribley, K. Zhang, and R. J. Kaufman, "Regulation of Apoptosis by the Unfolded Protein Response," in *Apoptosis: Methods and Protocols*, ed. P. Erhardt and A. Toth (Totowa, NJ: Humana Press, 2009), 191–204, https://doi.org/10.1007/978-1-60327-017-5_14.

110 **Eventually researchers discovered:** K. D. Wilkinson, "The Discovery of Ubiquitin-Dependent Proteolysis," *Proceedings of the National Academy of Sciences (PNAS) of the United States of America* 102, no. 43 (October 17, 2005): 15280–82, https://doi.org/10.1073/pnas.0504842102. There is a popular account of the discovery of the proteasome and the award of the Nobel Prize to Avram Hershko, Aaron Ciechanover, and Irwin Rose in "Popular Information: The Nobel Prize in Chemistry 2004," Nobel Prize online, accessed July 4, 2023, https://www.nobelprize.org/prizes/chemistry/2004/popular-information/.

110 **Deliberately introducing defects:** I. Saez and D. Vilchez, "The Mechanistic Links Between Proteasome Activity, Aging and Age-Related Diseases," *Current Genomics* 15, no. 1 (February 15, 2014): 38–51, https://doi.org/10.2174/1389202915501140306113344.

112 **By isolating strains:** K. Takeshig et al., "Autophagy in Yeast Demonstrated with Proteinase-Deficient Mutants and Conditions for Its Induction," *Journal of Cell Biology* 119, no. 2 (October 1992): 301–11, https://doi.org/10.1083/jcb.119.2.301; M. Tsukada and Y. Ohsumi, "Isolation and Characterization of Autophagy-Defective Mutants of *Saccharomyces cerevisiae*," *FEBS Letters* 333, nos. 1/2 (October 25, 1993): 169–74, https://doi.org/10.1016/0014-5793(93)80398-e.

113 **It has so many essential functions:** For a very reader-friendly description of autophagy, see "The Nobel Prize in Physiology or Medicine 2016: Yoshinori Ohsumi," press release, Nobel Prize online, October 3, 2016, https://www.nobelprize.org/prizes/medicine/2016/press-release/.

114 **Integrated stress response or ISR:** Two reviews of the integrated stress response are Harding, H. P. et al., "An integrated stress response regulates amino acid metabolism and resistance to oxidative stress," *Molecular Cell* 11, no. 3 (March 2003): 619–33, https://doi.org/10.1016/s1097-2765(03)00105-9; and Pakos-Zebrucka, K. et al. "The integrated stress response," *EMBO Reports* 17, no.10 (2016): 1374–95, https://doi.org/10.15252/embr.201642195. Its discovery in amino acid starvation is described in Dever, T. E. et al., "Phosphorylation of initiation factor 2 alpha by protein kinase GCN2 mediates gene-specific translational control of GCN4 in yeast," *Cell*

68. no. 3 (February 1992): 585–96, https://doi.org/10.1016/0092
-8674(92)90193-g and that in the unfolded protein response in Harding,
H. P. et al., "PERK is essential for translational regulation and cell
survival during the unfolded protein response," *Molecular Cell* 5, no. 5
(May 2000): 897-904, https://doi.org/10.1016/s1097-2765(00)80330-5.

114 If you delete the genes: M. Delépine et al., "*EIF2AK3*, Encoding
Translation Initiation Factor 2-Alpha Kinase 3, Is Mutated in Patients
with Wolcott-Rallison Syndrome," *Nature Genetics* 25, no. 4 (August
2000): 406–9, https://doi.org/10.1038/78085; H. P. Harding et al.,
"Diabetes Mellitus and Exocrine Pancreatic Dysfunction in Perk-/-
Mice Reveals a Role for Translational Control in Secretory Cell
Survival," *Molecular Cell* 7, no. 6 (June 2001): 1153–63, https://doi
.org/10.1016/s1097-2765(01)00264-7.

114 They also extend life span: S. J. Marciniak et al., "CHOP Induces
Death by Promoting Protein Synthesis and Oxidation in the Stressed
Endoplasmic Reticulum," *Genes & Development* 18, no. 24 (December
15, 2004): 3066–77, https://doi.org/10.1101/gad.1250704; M. D'Antonio et al., "Resetting Translational Homeostasis Restores Myelin-
ation in Charcot-Marie-Tooth Disease Type 1B Mice," *Journal of
Experimental Medicine* 210, no. 4 (April 8, 2013): 821–38, https://doi
.org/10.1084/jem.20122005; P. Tsaytler et al., "Selective Inhibition of
a Regulatory Subunit of Protein Phosphatase 1 Restores Proteostasis,"
Science 332, no. 6025 (April 1, 2011): 91–94, https://doi.org/10.1126
/science.1201396; H. Q. Jiang et al., "Guanabenz Delays the Onset of
Disease Symptoms, Extends Lifespan, Improves Motor Performance
and Attenuates Motor Neuron Loss in the SOD1 G93A Mouse
Model of Amyotrophic Lateral Sclerosis," *Neuroscience* 277 (March
2014): 132–38, https://doi.org/10.1016/j.neuroscience.2014.03.047;
I. Das et al., "Preventing Proteostasis Diseases by Selective Inhibition
of a Phosphatase Regulatory Subunit," *Science* 348, no. 6231 (April 10,
2015): 239–42, https://doi.org/10.1126/science.aaa4484.

114 whether they even affected ISR directly: A. Crespillo-Casado et al.,
"PPP1R15A-Mediated Dephosphorylation of eIF2α Is Unaffected by
Sephin1 or Guanabenz," *eLife* 6 (April 27, 2017): e26109, https://doi
.org/10.7554/eLife.26109.

114 According to their studies, deleting the genes: T. Ma et al., "Suppression of eIF2α Kinases Alleviates Alzheimer's Disease–Related
Plasticity and Memory Deficits," *Nature Neuroscience* 16, no. 9 (September 2013): 1299–305, https://doi.org/10.1038/nn.3486.

114 Even more surprisingly: Adam Piore, "The Miracle Molecule That

Could Treat Brain Injuries and Boost Your Fading Memory," *MIT Technology Review* 124, no. 5 (September/October 2021): https://www.technologyreview.com/2021/08/25/1031783/isrib-molecule-treat-brain-injuries-memory/; C. Sidrauski et al., "Pharmacological Brake-Release of mRNA Translation Enhances Cognitive Memory," *eLife* 2 (2013): e00498,https://doi.org/10.7554/eLife.00498; C. Sidrauski et al., "The Small Molecule ISRIB Reverses the Effects of Eif2α Phosphorylation on Translation and Stress Granule Assembly," *eLife* 4 (2015): e05033, https://doi.org/10.7554/eLife.05033; A. Chou et al., "Inhibition of the Integrated Stress Response Reverses Cognitive Deficits After Traumatic Brain Injury," *Proceedings of the National Academy of Sciences (PNAS) of the United States of America* 114, no. 31 (July 10, 2017): E6420–E6426, https://doi.org/10.1073/pnas.1707661114.

115 **Nahum Sonenberg:** Nahum Sonenberg, email message to the author, January 12, 2023.

116 **The key person:** D. M. Asher with M. A. Oldstone, *Carleton Gajdusek, 1923–2008: Biographical Memoirs* (Washington, DC: US National Academy of Sciences, 2013), http://www.nasonline.org/publications/biographical-memoirs/memoir-pdfs/gajdusek-d-carleton.pdf; Caroline Richmond, "Obituary: Carleton Gajdusek," *Guardian* (US edition) online, last modified February 25, 2009, https://www.theguardian.com/science/2009/feb/25/carleton-gajdusek-obituary.

117 **On the strength of this:** Frank Macfarlane Burnet studied how the immune system distinguishes between our own cells and foreign invaders and shared the 1960 Nobel Prize with Peter Medawar.

117 **"had an intelligence quotient":** Jay Ingram, *Fatal Flaws: How a Misfolded Protein Baffled Scientists and Changed the Way We Look at the Brain* (New Haven, CT: Yale University Press, 2013), as quoted in M. Goedert, "M. Prions and the Like," *Brain* 137, no. 1 (January 2014): 301–5, https://doi.org/10.1093/brain/awt179. See also J. Farquhar and D. C. Gajdusek, eds., *Early Letters and Field-Notes from the Collection of D. Carleton Gajdusek* (New York: Raven Press, 1981).

117 **This was a recent practice among the Fore:** J. Goodfield, "Cannibalism and Kuru," *Nature* 387 (June 26, 1997): 841, https://doi.org/10.1038/43043; R. Rhodes, "Gourmet Cannibalism in New Guinea Tribe," *Nature* 389 (September 4, 1997): 11, https://doi.org/10.1038/37853.

118 **He showed no remorse:** Ivin Molotsky, "Nobel Scientist Pleads Guilty to Abusing Boy," *New York Times* online, February 19, 1997, https://www.nytimes.com/1997/02/19/us/nobel-scientist-pleads

-guilty-to-abusing-boy.html. Two articles shed light on the sociology of Gajdusek's extended family: C. Spark, "Family Man: The Papua New Guinean Children of D. Carleton Gajdusek," *Oceania* 77, no. 3 (November 2007): 355–69, and C. Spark, "Carleton's Kids: The Papua New Guinean Children of D. Carleton Gajdusek," *Journal of Pacific History* 44, no. 1 (June 2009): 1–19.

120 The result is that the misfolded form: S. B. Prusiner, "Prions," *Proceedings of the National Academy of Sciences (PNAS) of the United States of America* 95, no. 23 (November 10, 1998): 13363–83, https://doi .org/10.1073/pnas.95.23.13363.

120 Alzheimer himself autopsied: A good review of the beta-amyloid hypothesis is R. E. Tanzi and L. Bertram, "Twenty Years of the Alzheimer's Disease Amyloid Hypothesis: A Genetic Perspective," *Cell* 120, no. 4 (February 25, 2005): 545–55, https://doi.org/10.1016 /j.cell.2005.02.008.

120 In 1984, scientists identified: G. G. Glenner and C. W. Wong, "Alzheimer's Disease and Down's Syndrome: Sharing of a Unique Cerebrovascular Amyloid Fibril Protein," *Biochemical and Biophysical Research Communications* 122, no. 3 (August 16, 1984): 1131–35, https://doi.org/10.1016/0006-291x(84)91209-9.

120 They turn out to have mutations: A. Goate et al., "Segregation of a Missense Mutation in the Amyloid Precursor Protein Gene with Familial Alzheimer's Disease," *Nature* 349, no. 6311 (February 21, 1991): 704–6, https://doi.org/10.1038/349704a0; M. C. Chartier-Harlin et al., "Early-Onset Alzheimer's Disease Caused by Mutations at Codon 717 of the Beta-amyloid Precursor Protein Gene," *Nature* 353, no. 6347 (October 31, 1991): 844–46, https://doi.org/10.1038/353844a0.

121 Perhaps these tau filaments: Jebelli, *In Pursuit of Memory*.

121 Although scientists were skeptical at first: P. Poorkaj et al., "Tau Is a Candidate Gene for Chromosome 17 Frontotemporal Dementia," *Annals of Neurology* 43, no. 6 (June 1998): 815–25, https://doi. org/10.1002/ana.410430617; M. Hutton et al., "Association of Missense and 5′-splice-site Mutations in *Tau* with the Inherited Dementia FTDP-17," *Nature* 393, no. 6686 (June 18, 1998): 702–5, https:// doi.org/10.1038/31508; M. G. Spillantini et al., "Mutation in the Tau Gene in Familial Multiple System Tauopathy with Presenile Dementia," *Proceedings of the National Academy of Sciences (PNAS) of the United States of America* 95, no. 13 (June 23, 1998): 7737–41, https://doi.org /10.1073/pnas.95.13.7737.

121 Rather, the aberrant: S. H. Scheres et al., "M. Cryo-EM Structures of Tau Filaments," *Current Opinion in Structural Biology* 64, 17–25 (2020). https://doi.org/10.1016/j.sbi.2020.05.011; M. Schweighauser et al., "Structures of α-synuclein Filaments from Multiple System Atrophy," *Nature* 585, no. 7825 (September 2020): 464–69, https://doi.org/10.1038/s41586-020-2317-6; Y. Yang et al., "Cryo-EM Structures of Amyloid-β 42 Filaments from Human Brains," *Science* 375, no. 6577 (January 13, 2022): 167–72, https://doi.org/10.1126/science.abm7285.

121 We do know that if you delete the genes: H. Zheng et al., "Beta-Amyloid Precursor Protein-Deficient Mice Show Reactive Gliosis and Decreased Locomotor Activity," *Cell* 81, no. 4 (May 19, 1995): 525–31, https://doi.org/10.1016/0092-8674(95)90073-x.

122 There is a growing feeling: M. Goedert, M. Masuda-Suzukake, and B. Falcon, "Like Prions: The Propagation of Aggregated Tau and α-synuclein in Neurodegeneration," *Brain* 140, no. 2 (February 2017): 266–78, https://doi.org/10.1093/brain/aww230; A. Aoyagi et al., "Aβ and Tau Prion-like Activities Decline with Longevity in the Alzheimer's Disease Human Brain," *Science Translational Medicine* 11, no. 490 (May 1, 2019): eaat8462, https://doi.org/10.1126/scitranslmed.aat8462; M. Jucker and L. C. Walker, "Self-propagation of Pathogenic Protein Aggregates in Neurodegenerative Diseases," *Nature* 501, no. 7465 (September 5, 2013): 45–51, https://doi.org/10.1038/nature12481.

123 Very recently, therapies: C. H. van Dyck et al., "Lecanemab in Early Alzheimer's Disease," *New England Journal of Medicine* 388, no. 1 (January 5, 2023): 9–21, https://doi.org/10.1056/nejmoa2212948; M. A. Mintun et al, "Donanemab in Early Alzheimer's Disease," *New England Journal of Medicine* 384 (May 6, 2021): 1691–1704, https://doi.org/ 10.1056/NEJMoa2100708. See also the more recent discussion by S. Reardon, "Alzheimer's Drug Donanemab: What Promising Trial Means for Treatments," *Nature* 617 (May 4, 2023): 232–33, https://doi.org/10.1038/d41586-023-01537-5.

7. Less Is More

125 Now, in a time of plenty: J. V. Neel, "Diabetes Mellitus: A 'Thrifty' Genotype Rendered Detrimental by 'Progress,'" *American Journal of Human Genetics* 14, no. 4 (December 1962): 353–62, https://www.ncbi.nlm.nih.gov/pmc/articles/PMC1932342/.

125 **"drifty genes":** J. R. Speakman, "Thrifty Genes for Obesity and the Metabolic Syndrome—Time to Call off the Search?," *Diabetes and Vascular Disease Research* 3, no. 1 (May 2006): 7–11, https://doi.org /10.3132/dvdr.2006.010; J. R. Speakman, "Evolutionary Perspectives on the Obesity Epidemic: Adaptive, Maladaptive, and Neutral Viewpoints," *Annual Review of Nutrition* 33, no. 1 (July 2013): 289–317, https://doi.org/10.1146/annurev-nutr-071811-150711.

125 **The first studies to test this:** Two surveys of the field from the mid-2000s are E. J. Masoro, "Overview of Caloric Restriction and Ageing," *Mechanisms of Ageing and Development* 126, no. 9 (September 2005): 913–22, https://doi.org/10.1016/j.mad.2005.03.012, and B. K. Kennedy, K. K. Steffen, and M. Kaeberlein, "Ruminations on Dietary Restriction and Aging," *Cellular and Molecular Life Sciences* 64, no. 11 (June 2007): 1323–28, doi: 10.1007/s00018-007-6470-y.

126 **Moreover, they appeared to have delayed:** R. Weindruch and R. L. Walford, *The Retardation of Aging and Disease by Dietary Restriction* (Springfield, IL: C. C. Thomas, 1988), as quoted in Kennedy, Steffen, and Kaeberlein, "Ruminations," 1323–28; L. Fontana and L. Partridge, "Promoting Health and Longevity Through Diet: From Model Organisms to Humans," *Cell* 161, no. 1 (March 26, 2015): 106–18, https://doi.org/10.1016/j.cell.2015.02.020.

126 **In 2009:** R. J. Colman et al., "Caloric Restriction Delays Disease Onset and Mortality in Rhesus Monkeys," *Science* 325, no. 5937 (July 10, 2009): 201–4, https://doi.org/10.1126/science.1173635.

126 **But this was contradicted:** J. A. Mattison et al., "Impact of Caloric Restriction on Health and Survival in Rhesus Monkeys from the NIA Study," *Nature* 489, no. 7415 (September 13, 2012): 318–21, https:// doi.org/10.1038/nature11432. See the accompanying commentary by S. N. Austad, "Aging: Mixed Results for Dieting Monkeys," *Nature* 489, no. 7415 (September 13, 2012): 210–11, https://doi.org/10.1038 /nature11484, and a related news article in the same journal, A. Maxmen, "Calorie Restriction Falters in the Long Run," *Nature* 488, no. 7413 (August 30, 2012), 569, https://doi.org/10.1038/488569a.

127 **Any evidence for the effect of CR:** Laura A. Cassiday, "The Curious Case of Caloric Restriction," *Chemical & Engineering News* online, last modified August 3, 2009, https://cen.acs.org/articles/87/i31/Curious -Case-Caloric-Restriction.html.

127 **There is 5:2 fasting:** Gideon Meyerowitz-Katz, "Intermittent Fasting Is Incredibly Popular. But Is It Any Better Than Other Diets?," *Guardian* (US edition) online, last modified January 1, 2020, https://

www.theguardian.com/commentisfree/2020/jan/02/intermittent
-fasting-is-incredibly-popular-but-is-it-any-better-than-other-diets.

127 **They concluded that matching:** V. Acosta-Rodríguez et al., "Circadian Alignment of Early Onset Caloric Restriction Promotes Longevity in Male C57BL/6J Mice," *Science* 376, no. 6598 (May 5, 2022): 1192–202, https://doi.org/10.1126/science.abk0297. See the accompanying commentary in S. Deota and S. Panda, "Aligning Mealtimes to Live Longer," *Science* 376, no. 6598 (May 5, 2022): 1159–60, https://doi.org/10.1126/science.adc8824.

128 **In particular, sleep deprivation:** Matthew Walker, *Why We Sleep: The New Science of Sleep and Dreams* (New York: Scribner, 2017). See in particular chapter 8 for its effects on aging.

128 **According to a recent study:** A. Vaccaro et al., "Sleep Loss Can Cause Death Through Accumulation of Reactive Oxygen Species in the Gut," *Cell* 181, no. 6 (June 11, 2020): 1307–28.e15, https://doi.org/10.1016/j.cell.2020.04.049. See also a popular discussion of this in Veronique Greenwood, "Why Sleep Deprivation Kills," Quanta, last modified June 4, 2020, https://www.quantamagazine.org/why-sleep-deprivation-kills-20200604/, and Steven Strogatz, "Why Do We Die Without Sleep?," *The Joy of Why* (podcast, transcription), March 22, 2022, https://www.quantamagazine.org/why-do-we-die-without-sleep-20220322/.

128 **In one study:** C.-Y Liao et al., "Genetic Variation in Murine Lifespan Response to Dietary Restriction: From Life Extension to Life Shortening," *Aging Cell* 9, no. 1 (February 2010): 92–95, https://doi.org/10.1111/j.1474-9726.2009.00533.x.

128 **He felt that animals:** L. Hayflick, "Dietary Restriction: Theory Fails to Satiate," *Science* 329, no. 5995 (August 27, 2010): 1014, https://www.science.org/doi/10.1126/science.329.5995.1014; L. Fontana, L. Partridge, and V. Longo, "Dietary Restriction: Theory Fails to Satiate—Response," *Science* 329, no. 5995 (August 27, 2010): 1015, https://www.science.org/doi/10.1126/science.329.5995.1015.

128 **Moreover, when scientists:** Saima May Sidik, "Dietary Restriction Works in Lab Animals, But It Might Not Work in the Wild," *Scientific American* online, last modified December 20, 2022, https://www.scientificamerican.com/article/dietary-restriction-works-in-lab-animals-but-it-might-not-work-in-the-wild/.

129 **On a more granular level:** Fontana and Partridge, "Promoting Health and Longevity," 106–18.

129 **Among its other reported downsides:** J. R. Speakman and S. E.

Mitchell, "Caloric Restriction," *Molecular Aspects of Medicine* 32, no. 3 (June 2011): 159–221, https://doi.org/10.1016/j.mam.2011.07.001.

130 **In 1964:** For an intriguing history of the discovery of rapamycin, see Bethany Halford, "Rapamycin's Secrets Unearthed," *Chemical & Engineering News* online, last modified July 18, 2016, https://cen.acs .org/articles/94/i29/Rapamycins-Secrets-Unearthed.html, which is the basis for the next few paragraphs. See also David Stipp, "A New Path to Longevity," *Scientific American* online, last modified January 1, 2012), https://www.scientificamerican.com/article/a-new-path-to -longevity/.

131 **Here our story shifts to Basel, Switzerland:** U. S. Neill, "A Conversation with Michael Hall," *Journal of Clinical Investigation* 127, no. 11 (November 1, 2017): 3916–17, https://doi.org/10.1172/jci97760; C. L. Williams, "Talking TOR: A Conversation with Joe Heitman and Rao Movva," *JCI Insight* 3, no. 4 (February 22, 2018): e99816, https:// doi.org/10.1172/jci.insight.99816.

136 **How cell size and shape are controlled:** M. B. Ginzberg, R. Kafri, and M. Kirschner, "On Being the Right (Cell) Size," *Science* 348, no. 6236 (May 15, 2015): 1245075, https://doi.org/10.1126/science .1245075.

136 **His paper was rejected:** N. C. Barbet et al., "TOR Controls Translation Initiation and Early G1 Progression in Yeast," *Molecular Biology of the Cell* 7, no. 1 (January 1, 1996): 25–42, https://doi.org/10.1091 /mbc.7.1.25. For Hall's recollections about the early days and the difficulty of getting the scientific community to accept that cell growth was actively controlled, see M. N. Hall, "TOR and Paradigm Change: Cell Growth Is Controlled," *Molecular Biology of the Cell* 27, no. 18 (September 15, 2016): 2804–6, https://doi.org/10.1091/mbc.E15-05 -0311.

138 **We can now see:** D. Papadopoli et al., "mTOR as a Central Regulator of Lifespan and Aging," *F1000 Research* 8 (July 2, 2019): 998, https://doi.org/10.12688/f1000research.17196.1; G. Y. Liu and D. M. Sabatini, "mTOR at the Nexus of Nutrition, Growth, Ageing and Disease," *Nature Reviews Molecular Biology* 21, no. 4 (April 2020): 183–203, https://doi.org/10.1038/s41580-019-0199-y.

139 **It turns out that both a defective TOR:** L. Partridge, M. Fuentealba, and B. K. Kennedy, "The Quest to Slow Ageing Through Drug Discovery," *Nature Reviews Drug Discovery* 19, no. 8 (August 2020): 513–32, https://doi.org/10.1038/s41573-020-0067-7.

139 **Strikingly, even short courses:** D. E. Harrison et al., "Rapamycin Fed

Late in Life Extends Lifespan in Genetically Heterogeneous Mice," *Nature* 460, no. 7253 (July 16, 2009): 392–95, https://doi.org/10.1038/nature08221; see the accompanying commentary by M. Kaeberlein and R. K. Kennedy, "Ageing: A Midlife Longevity Drug?," *Nature* 460, no. 7253 (July 16, 2009): 331–32, https://doi.org/10.1038/460331a.

139 **Rapamycin also delayed:** F. M. Menzies and D. C. Rubinsztein, "Broadening the Therapeutic Scope for Rapamycin Treatment," *Autophagy* 6, no. 2 (February 2010): 286–87, https://doi.org/10.4161/auto.6.2.11078.

139 **While rapamycin inhibits:** K. Araki et al., "mTOR Regulates Memory CD8 T-cell Differentiation," *Nature* 460, no. 7251 (July 2, 2009): 108–12, https://doi.org/10.1038/nature08155.

140 **Another study, from 2009, showed that administering rapamycin:** C. Chen et al. "mTOR Regulation and Therapeutic Rejuvenation of Aging Hematopoietic Stem Cells," *Science Signaling* 2, no. 98 (November 24, 2009): ra75, https://doi.org/10.1126/scisignal.2000559.

140 **As one might expect:** A. M. Eiden, "Molecular Pathways: Increased Susceptibility to Infection Is a Complication of mTOR Inhibitor Use in Cancer Therapy," *Clinical Cancer Research* 22, no. 2 (January 15, 2016): 277–83, https://doi.org/10.1158/1078-0432.ccr-14-3239.

140 **"warrants caution":** A. J. Pagán et al., "mTOR-Regulated Mitochondrial Metabolism Limits Mycobacterium-Induced Cytotoxicity, *Cell* 185, no. 20 (September 29, 2022): 3720–38, e13, https://doi.org/10.1016/j.cell.2022.08.018.

141 **"I suppose the rapamycin advocates":** Michael Hall, email message to the author, September 29, 2022.

141 **The consortium will analyze:** K. E. Creevy et al., "An Open Science Study of Ageing in Companion Dogs," *Nature* 602, no. 7895 (February 2022): 51–57, https://doi.org/10.1038/s41586-021-04282-9.

142 **They go on to suggest:** M. V. Blagosklonny and M. N. Hall, "Growth and Aging: A Common Molecular Mechanism," *Aging* 1, no. 4 (April 20, 2009): 357–62, https://doi.org/10.18632/aging.100040.

8. Lessons from a Lowly Worm

143 **A study of 2,700 Danish twins:** A. M. Herskind et al., "The Heritability of Human Longevity: A Population-Based Study of 2,872 Danish Twin Pairs Born 1870–1900," *Human Genetics* 97, no. 3 (March 1996): 319–23, https://doi.org/10.1007/BF02185763.

143 **Once he and Crick:** Their views and plans are outlined in a 1971 report by Francis Crick and Sydney Brenner. See F. H. C. Crick and S. Brenner, *Report to the Medical Research Council: On the Work of the Division of Molecular Genetics, Now the Division of Cell Biology, from 1961–1971* (Cambridge, UK: MRC Laboratory of Molecular Biology, November 1971), https://profiles.nlm.nih.gov/spotlight/sc/catalog/nlm:nlmuid-101584582X71-doc.

144 **Scientists went on to identify:** For this work, Brenner was awarded the 2002 Nobel Prize in Physiology or Medicine, along with two of his former colleagues, John Sulston and Robert Horvitz. "The Nobel Prize in Physiology or Medicine 2002," Nobel Prize online, accessed July 22, 2023, https://www.nobelprize.org/prizes/medicine/2002/summary/.

145 **As Hirsh recalled:** David Hirsh, email message to the author, August 1, 2022.

146 **Instead, it turned out:** D. B. Friedman and T. E. Johnson, "A Mutation in the *age-1* Gene in *Caenorhabditis elegans* Lengthens Life and Reduces Hermaphrodite Fertility," *Genetics* 118, no. 1 (January 1, 1988): 75–86, https://doi.org/10.1093/genetics/118.1.75.

146 **Johnson went on to show:** T. E. Johnson, "Increased Life-Span of *age-1* Mutants in *Caenorhabditis elegans* and Lower Gompertz Rate of Aging," *Science* 249, no. 4971 (August 24, 1990): 908–12, https://doi.org/10.1126/science.2392681.

146 **Even after it finally appeared in the prestigious journal *Science* in 1990:** David Stipp's book *The Youth Pill: Scientists at the Brink of an Anti-Aging Revolution* (New York: Penguin, 2010) contains an engaging and detailed account of the history, personalities, and science behind the discovery of aging mutants.

147 **she felt inspired:** Two firsthand accounts by Kenyon and Johnson of their discoveries are C. Kenyon, "The First Long-Lived Mutants: Discovery of the Insulin/IGF-1 Pathway for Ageing," *Philosophical Transactions of the Royal Society B: Biological Sciences* 366, no. 1561 (January 12, 2001): 9–16, https://doi.org/10.1098/rstb.2010.0276, and T. E. Johnson, "25 Years After *age-1*: Genes, Interventions and the Revolution in Aging Research," *Experimental Gerontology* 48, no. 7 (July 2013): 640–43, https://doi.org/10.1016/j.exger.2013.02.023.

147 **her 1993 paper:** C. Kenyon et al., "*A C. elegans* Mutant That Lives Twice as Long as Wild Type," *Nature* 366, no. 6454 (December 2, 1993): 461–64, https://doi.org/10.1038/366461a0.

147 **Apart from her stellar academic pedigree:** Stipp, *Youth Pill*.

148 **"I thought, 'Oh, gosh'":** Ibid.

149 **As it turns out, the *age-1* gene originally identified:** The key papers for the identity of some of the key genes are (*daf-2*) K. D. Kimura, H. A. Tissenbaum, and G. Ruvkun, "*daf-2*, an Insulin Receptor-Like Gene That Regulates Longevity and Diapause in *Caenorhabditis elegans*," *Science* 277, no. 5328 (August 15, 1997): 942–46, https://doi.org/10.1126/science.277.5328.942; (*age-1*, which turned out to be the same as *daf-23*), J. Z. Morris, H. A. Tissenbaum, and G. Ruvkun, "A Phosphatidylinositol-3-OH Kinase Family Member Regulating Longevity and Diapause in *Caenorhabditis elegans*, *Nature* 382, no. 6591 (August 8, 1996): 536–39, https://doi.org/10.1038/382536a0; (daf-16), S. Ogg et al., "The Fork Head Transcription Factor DAF-16 Transduces Insulin-like Metabolic and Longevity Signals in *C. elegans*," *Nature* 389, no. 6654 (October 30, 1997): 994–99, https://doi.org/10.1038/40194, and K. Lin et al., "*daf-16*: An HNF-3/Forkhead Family Member That Can Function to Double the Life-Span of *Caenorhabditis elegans*," *Science* 278, no. 5341 (November 14, 1997): 1319–22, https://doi.org/10.1126/science .278.5341.1319.

149 **"constitute a treasure trove":** C. J. Kenyon, "The Genetics of Ageing," *Nature* 464, no. 7288 (March 25, 2010): 504–12, https://doi .org/10.1038/nature08980.

150 **Among the many reasons for this:** H. Yan et al., "Insulin Signaling in the Long-Lived Reproductive Caste of Ants," *Science* 377, no. 6610 (September 1, 2022): 1092–99, https://doi.org/10.1126/science .abm8767.

150 **However, if the experiment is repeated—but this time using a strain:** E. Cohen et al., "Opposing Activities Protect Against Age-Onset Proteotoxicity," *Science* 313, no. 5793 (September 15, 2006): 1604–10, https://doi.org/10.1126/science.1124646.

150 **Deleting the gene that codes for a protein:** D. J. Clancy et al., "Extension of Life-span by Loss of CHICO, a *Drosophila* Insulin Receptor Substrate Protein," *Science* 292, no. 5514 (April 6, 2001): 104–6, https://doi.org/10.1126/science.1057991.

150 **The IGF-1 receptor is essential:** M. Holzenberger et al., "IGF-1 Receptor Regulates Lifespan and Resistance to Oxidative Stress in Mice," *Nature* 421, no. 6919 (January 9, 2003): 182–87, https://doi.org /10.1038/nature01298; G. J. Lithgow and M. S. Gill, "Physiology: Cost-Free Longevity in Mice," *Nature* 421, no. 6919 (January 9, 2003): 125–26, https://doi.org/10.1038/421125a.

151 **An analysis of subjects:** D. A. Bulger et al., "*Caenorhabditis ele-*

gans DAF-2 as a Model for Human Insulin Receptoropathies," *G3 Genes | Genomes | Genetics* 7, no. 1 (January 1, 2017): 257–68, https://doi.org/10.1534/g3.116.037184.

151 Mutations known to impair IGF-1: Y. Suh et al., "Functionally Significant Insulin-like Growth Factor I Receptor Mutations in Centenarians," *Proceedings of the National Academy of Sciences (PNAS) of the United States of America* 105, no. 9 (March 4, 2008): 3438–42, https://doi.org/10.1073/pnas.0705467105; T. Kojima et al., "Association Analysis Between Longevity in the Japanese Population and Polymorphic Variants of Genes Involved in Insulin and Insulin-like Growth Factor 1 Signaling Pathways," *Experimental Gerontology* 39, nos. 11/12 (November/December 2004): 1595–98, https://doi.org/10.1016/j.exger.2004.05.007.

151 Variants in proteins: See references in Kenyon, "Genetics of Ageing," 504–12.

151 Exactly as you might predict: S. Honjoh et al., "Signalling Through RHEB-1 Mediates Intermittent Fasting-Induced Longevity in *C. elegans*," *Nature* 457, no. 7230 (February 5, 2009): 726–30, https://doi.org/10.1038/nature07583.

152 This means that caloric restriction: B. Lakowski and S. Hekimi, "The Genetics of Caloric Restriction in *Caenorhabditis elegans*," *Proceedings of the National Academy of Sciences (PNAS) of the United States of America* 95, no. 22 (October 27, 1998): 13091–96, https://doi.org/10.1073/pnas.95.22.13091.

152 When worms were subjected: D. W. Walker et al., "Evolution of Lifespan in *C. elegans*," *Nature* 405, no. 6784 (May 18, 2000): 296–97, https://doi.org/10.1038/35012693.

152 To understand the difference: Stephen O'Rahilly, conversation with the author, August 11, 2022.

154 Because of recent advances: H. R. Bridges et al., "Structural Basis of Mammalian Respiratory Complex I Inhibition by Medicinal Biguanides," *Science* 379, no. 6630 (January 26, 2023): 351–57, https://www.science.org/doi/10.1126/science.ade3332.

155 Disrupting our ability to utilize glucose: G. Rena, D. G. Hardie, and E. R. Pearson, "The Mechanisms of Action of Metformin," *Diabetologia* 60, no. 9 (September 2017): 1577–85, https://doi.org/10.1007/s00125-017-4342-z; T. E. LaMoia and G. I. Shulman, "Cellular and Molecular Mechanisms of Metformin Action," *Endocrine Reviews* 42, no. 1 (February 2021): 77–96, https://doi.org/10.1210/endrev/bnaa023.

155 **Although some studies have claimed:** L. C. Gormsen et al., "Metformin Increases Endogenous Glucose Production in Non-Diabetic Individuals and Individuals with Recent-Onset Type 2 Diabetes," *Diabetologia* 62, no. 7 (July 2019): 1251–56, https://doi.org/10.1007/s00125-019-4872-7.

155 **According to another study, the drug alters:** H. Wu et al., "Metformin Alters the Gut Microbiome of Individuals with Treatment-Naive Type 2 Diabetes, Contributing to the Therapeutic Effects of the Drug," *Nature Medicine* 23, no. 7 (July 2017): 850–58, https://doi.org/10.1038/nm.4345.

155 **Steve O'Rahilly's work demonstrates:** A. P. Coll et al., "GDF15 Mediates the Effects of Metformin on Body Weight and Energy Balance," *Nature* 578, no. 7795 (February 2020): 444–48, https://doi.org/10.1038/s41586-019-1911-y.

155 **In the first, from the National Institute on Aging, long-term treatment:** A. Martin-Montalvo et al., "Metformin Improves Healthspan and Lifespan in Mice," *Nature Communications* 4 (2013): 2192, https://doi.org/10.1038/ncomms3192.

155 **A second study, in humans:** C. A. Bannister et al., "Can People with Type 2 Diabetes Live Longer Than Those Without? A Comparison of Mortality in People Initiated with Metformin or Sulphonylurea Monotherapy and Matched, Non-Diabetic Controls," *Diabetes, Obesity and Metabolism* 16, no. 11 (November 2014): 1165–73, https://doi.org/10.1111/dom.12354.

155 **One, from 2016, concluded that metformin:** M. Claesen et al., "Mortality in Individuals Treated with Glucose-Lowering Agents: A Large, Controlled Cohort Study," *Journal of Clinical Endocrinology & Metabolism* 101, no. 2 (February 1, 2016): 461–69, https://doi.org/10.1210/jc.2015-3184.

156 **Curiously, some of the toxicity:** L. Espada et al., "Loss of Metabolic Plasticity Underlies Metformin Toxicity in Aged *Caenorhabditis Elegans*," *Nature Metabolism* 2, no. 11 (November 2020): 1316–31, https://doi.org/10.1038/s42255-020-00307-1.

156 **Metformin also undermined:** A. R. Konopka et al., "Metformin Inhibits Mitochondrial Adaptations to Aerobic Exercise Training in Older Adults," *Aging Cell* 18, no. 1 (February 2019): e12880, https://doi.org/10.1111/acel.12880.

156 **And one study claimed that diabetics:** Y. C. Kuan et al., "Effects of Metformin Exposure on Neurodegenerative Diseases in Elderly

Patients with Type 2 Diabetes Mellitus," *Progress in Neuropsychopharmacol and Biological Psychiatry* 79, pt. B (October 3, 2017): 1777–83 (2017), https://doi.org/10.1016/j.pnpbp.2017.06.002.

156 **The study's goal is to see:** "The Tame Trial: Targeting the Biology of Aging: Ushering a New Era of Interventions," American Federation for Aging Research (AFAR) online, accessed August 1, 2023, https://www.afar.org/tame-trial.

158 **That was exactly the skepticism:** A detailed account of how Guarente became involved in this research and his laboratory's early discoveries is found in his book, Lenny Guarente, *Ageless Quest: One Scientist's Search for Genes That Prolong Youth* (Cold Spring Harbor, NY: Cold Spring Harbor Press, 2003).

158 **Increasing the amount of Sir2:** M. Kaeberlein, M. McVey, and L. Guarente, "The SIR2/3/4 Complex and SIR2 Alone Promote Longevity in *Saccharomyces cerevisiae* by Two Different Mechanisms," *Genes and Development* 13, no. 19, October 1, 1994, 2570–80, https://doi.org/10.1101/gad.13.19.2570.

158 **They soon found, with mounting excitement:** B. Rogina and S. L. Helfand, "Sir2 Mediates Longevity in the Fly Through a Pathway Related to Calorie Restriction," *Proceedings of the National Academy of Sciences (PNAS) of the United States of America* 101, no. 45 (November 2004): 15998–6003, https://doi.org/10.1073/Pnas.040418410; H. A. Tissenbaum and L. Guarente, "Increased Dosage of a Sir-2 Gene Extends Lifespan in *Caenorhabditis Elegans*," *Nature* 410, no. 6825 (March 8, 2001): 227–30, https://doi.org/10.1038/35065638.

159 **Sir2 turns out to be a deacetylase:** S. Imai et al., "Transcriptional Silencing and Longevity Protein Sir2 Is an NAD-Dependent Histone Deacetylase," *Nature* 403, no. 6771 (February 17, 2000): 795–800, https://doi.org/10.1038/35001622; W. Dang et al., "Histone H4 Lysine 16 Acetylation Regulates Cellular Lifespan," *Nature* 459, no. 7248 (June 11, 2009): 802–7, https://doi.org/10.1038/nature08085.

159 **Sure enough, in both flies and yeast:** S. J. Lin, P. A. Defossez, and L. Guarente, "Requirement of NAD and *SIR2* for Life-span Extension by Calorie Restriction in *Saccharomyces cerevisiae*," *Science* 289, no. 5487 (September 22, 2000): 2126–28, https://doi.org/10.1126/science.289.5487.2126; Rogina and Helfand, "Sir2 Mediates Longevity in the Fly," 15998–6003.

159 **"When single genes are changed":** L. Guarente and C. Kenyon,

"Genetic Pathways That Regulate Ageing in Model Organisms," *Nature* 408, no. 6809 (November 9, 2000): 255–62, https://doi.org/10.1038/35041700.

160 **Finally, here was scientific evidence:** K. T. Howitz. et al., "Small Molecule Activators of Sirtuins Extend *Saccharomyces cerevisiae* Lifespan," *Nature* 425, no. 6809 (November 9, 2000): 191–96, https://doi.org/10.1038/nature01960.

160 **Although the mice remained overweight:** J. A. Baur et al., "Resveratrol Improves Health and Survival of Mice on a High-Calorie Diet," *Nature* 444, no. 7117 (November 16, 2006): 337–42, https://doi.org/10.1038/nature05354; M. Lagouge et al., "Resveratrol Improves Mitochondrial Function and Protects Against Metabolic Disease by Activating SIRT1 and PGC-1alpha," *Cell* 127, no. 6 (December 15, 2006): 1109–22, https://doi.org/10.1016/j.cell.2006.11.013.

161 **Among other things:** M. Kaeberlein et al., "Sir2-Independent Life Span Extension by Calorie Restriction in Yeast," *PLoS Biology* 2, no. 9 (September 2004): E296, https://doi.org/10.1371/journal.pbio.0020296.

161 **Not only that, but they did not find:** M. Kaeberlein et al., "Substrate-Specific Activation of Sirtuins by Resveratrol," *Journal of Biological Chemistry* 280, no. 17 (April 2005): 17038–45, https://doi.org/10.1074/jbc.M500655200.

161 **Pharmaceutical companies do not usually:** M. Pacholec et al., "SRT1720, SRT2183, SRT1460, and Resveratrol Are Not Direct Activators of SIRT1," *Journal of Biological Chemistry* 285, no. 11 (March 2010): 8340–51, https://doi.org/10.1074/jbc.M109.088682.

162 **Five years after the sale:** John La Mattina, "Getting the Benefits of Red Wine from a Pill? Not Likely," *Forbes* online, last modified March 19, 2013, https://www.forbes.com/sites/johnlamattina/2013/03/19/getting-the-benefits-of-red-wine-from-a-pill-not-likely/.

162 **This led to another commentary:** B. P. Hubbard et al., "Evidence for a Common Mechanism of SIRT1 Regulation by Allosteric Activators," *Science* 339, no. 6124 (March 8, 2013): 1216–19, https://doi.org/10.1126/science.1231097; H. Yuan and R. Marmorstein, "Red Wine, Toast of the Town (Again)," *Science* 339, no. 6124 (March 8, 2013): 1156–57, https://doi.org/10.1126/science.1236463.

162 **None of them had any significant effect:** R. Strong et al., "Evaluation of Resveratrol, Green Tea Extract, Curcumin, Oxaloacetic Acid, and Medium-Chain Triglyceride Oil on Life Span of Genetically

Heterogeneous Mice," *Journals of Gerontology: Series A* 68, no. 1 (January 2013): 6–16, https://doi.org/10.1093/gerona/gls070.

162 **Sir2 activation actually reduces:** P. Fabrizio et al., "Sir2 Blocks Extreme Life-span Extension," *Cell* 123, no. 4 (November 18, 2005): 655–67, https://doi.org/10.1016/j.cell.2005.08.042; see also commentary by B. K. Kennedy, E. D. Smith, and M. Kaeberlein, "The Enigmatic Role of Sir2 in Aging," *Cell* 123, no. 4 (November 18, 2005): 548–50, https://doi.org/10.1016/j.cell.2005.11.002.

163 **Feeling embattled:** C. Burnett et al., "Absence of Effects of Sir2 Overexpression on Lifespan in *C. elegans* and *Drosophila*," *Nature* 477, no. 7365 (September 21, 2011): 482–85, https://doi.org/10.1038/nature10296; K. Baumann, "Ageing: A Midlife Crisis for Sirtuins," *Nature Reviews Molecular Cell Biology* 12, no. 11 (October 21, 2011): 688, https://doi.org/10.1038/nrm3218; D. B. Lombard et al., "Ageing: Longevity Hits a Roadblock," *Nature* 477, no. 7365 (September 21, 2011): 410–11, https://doi.org/10.1038/477410a; M. Viswanathan and L. Guarente, "Regulation of *Caenorhabditis elegans* lifespan by *sir-2.1* Transgenes," *Nature* 477, no. 7365 (September 21, 2011): E1–2, https://doi.org/10.1038/nature10440.

163 **The protein is also a histone:** R. Mostoslavsky et al., "Genomic Instability and Aging-like Phenotype in the Absence of Mammalian SIRT6," *Cell* 124, no. 2 (January 24, 2006): 315–29, https://doi.org/10.1016/j.cell.2005.11.044; E. Michishita et al. "SIRT6 Is a Histone H3 Lysine 9 Deacetylase That Modulates Telomeric Chromatin," *Nature* 452, no. 7186 (March 27, 2008): 492–96, https://doi.org/10.1038/nature06736; A. Roichman et al., "SIRT6 Overexpression Improves Various Aspects of Mouse Healthspan," *Journals of Gerontology: Series A* 72, no. 5 (May 1, 2017): 603–15, https://doi.org/10.1093/gerona/glw152; X. Tian et al., "SIRT6 Is Responsible for More Efficient DNA Double-Strand Break Repair in Long-Lived Species," *Cell* 177, no. 3 (April 18, 2019): 622–38.e22, https://doi.org/10.1016/j.cell.2019.03.043.

163 **Many in the gerontology community:** C. Brenner, "Sirtuins Are Not Conserved Longevity Genes," *Life Metabolism* 1, no. 2 (October 2022), 122–33, https://doi.org/10.1093/lifemeta/loac025.

164 **It is made by the body:** P. Belenky, K. L. Bogan, and C. Brenner, "NAD+ Metabolism in Health and Disease," *Trends in Biochemical Sciences* 32, no. 1 (January 2017): 12–19, https://doi.org/10.1016/j.tibs.2006.11.006.

164 **It can also cause a host:** H. Massudi et al., "Age-Associated Changes in Oxidative Stress and NAD+ Metabolism in Human Tissue," *PLoS One* 7, no. 7 (2012): e42357, https://doi.org/10.1371/journal. pone.0042357; X. H. Zhu et al., "In Vivo NAD Assay Reveals the Intracellular NAD Contents and Redox State in Healthy Human Brain and Their Age Dependences," *Proceedings of the National Academy of Sciences (PNAS) of the United States of America* 112, no. 9 (February 17, 2015): 2876–81, https://doi.org/10.1073/pnas.1417921112; A. J. Covarrubias et al., "NAD+ Metabolism and Its Roles in Cellular Processes During Ageing," *Nature Reviews Molecular Cell Biology* 22, no. 2 (February 2021): 119–41, https://doi.org/10.1038/s41580-020 -00313-x.

164 **Increasing NAD levels:** H. Zhang et al., "NAD+ Repletion Improves Mitochondrial and Stem Cell Function and Enhances Life Span in Mice," *Science* 352, no. 6292 (April 28, 2016): 1436–43, https://doi .org/10.1126/science.aaf2693; see also the commentary on this report by L. Guarente, "The Resurgence of NAD+," *Science* 352, no. 6292 (April 28, 2016): 1396–97, https://doi.org/10.1126/science.aag1718; K. F. Mills et al., "Long-Term Administration of Nicotinamide Mononucleotide Mitigates Age-Associated Physiological Decline in Mice," *Cell Metabolism* 24, no. 6 (December 13, 2016): 795–806, https://doi .org/10.1016/j.cmet.2016.09.013.

165 **"I expressly tell people":** Charles Brenner, email message to the author, January 22, 2023.

165 **The results of taking:** Partridge, Fuentealba, and Kennedy, "Quest to Slow Ageing," 513–32.

165 **Global sales of NMN:** Global News Wire, "Nicotinamide Mononucleotide (NMN) Market Will Turn Over USD 251.2 to Revenue to Cross USD 953 Million in 2022 to 2028 Research by Business Opportunities, Top Companies, Opportunities Planning, Market-Specific Challenges," August 19, 2022, https://www.globenewswire .com/en/news-release/2022/08/19/2501489/0/en/Nicotinamide-Mono nucleotide-NMN-Market-will-Turn-over-USD-251-2-to-Revenue -to-Cross-USD-953-million-in-2022-to-2028-Research-by-Business -Opportunities-Top-Companies-opportunities-p.html.

9. The Stowaway Within Us

167 **"I quit my job":** Martin Weil, "Lynn Margulis, Leading Evolutionary Biologist, Dies at 73," *Washington Post* online, November 26, 2011,

https://www.washingtonpost.com/local/obituaries/lynn-margulis
-leading-evolutionary-biologist-dies-at-73/2011/11/26/gIQAQ
5dezN_story.html.

167 **Margulis wrote an essay:** Lynn Margulis, "Two Hit, Three Down—
The Biggest Lie: David Ray Griffin's Work Exposing 9/11," in
Dorion Sagan, ed., *Lynn Margulis: The Life and Legacy of a Scientific
Rebel* (White River Junction, VT: Chelsea Green, 2012), 150–55.

167 **questioned whether the human immunodeficiency virus (HIV):**
Joanna Bybee, "No Subject Too Sacred," in Sagan, ed. *Lynn Margulis*,
156–62.

167 **You could think of Margulis's idea:** L. Sagan, "On the Origin of
Mitosing Cells," *Journal of Theoretical Biology 14*, no. 3 (March 14,
1967): 255–74, https://doi.org/10.1016/0022-5193(67)90079-3.

170 **In the same way that water:** The idea that ATP is made by using a
proton gradient across a membrane was proposed by Peter Mitchell
and highly controversial initially. He went on to receive the 1978 Nobel
Prize. See: Royal Swedish Academy of Sciences, "The Nobel Prize in
Chemistry 1978: Peter Mitchell," press release, October 17, 1978, available at Nobel Prize online, https://www.nobelprize.org/prizes/chemis
try/1978/press-release/. Part of the 1997 Chemistry Nobel Prize was
awarded to Paul Boyer and John Walker for their work on the molecular turbine that actually makes the ATP. The Nobel press release has an
excellent description of it: Royal Swedish Academy of Sciences, "The
Nobel Prize in Chemistry 1997: Paul D. Boyer, John E. Walker, Jens C.
Skou," press release, October 15, 1997, available at Nobel Prize online,
https://www.nobelprize.org/prizes/chemistry/1997/press-release/.

171 **The human body has to generate:** F. Du et al., "Tightly Coupled Brain
Activity and Cerebral ATP Metabolic Rate," *Proceedings of the National
Academy of Sciences (PNAS) of the United States of America* 105, no. 17
(April 29, 2008): 6409–14, https://doi.org/10.1073/pnas.0710766105.
For a popular account of this article, see N. Swaminathan, "Why Does
the Brain Need So Much Power?," *Scientific American* online, April 29,
2008, https://www.scientificamerican.com/article/why-does-the-brain
-need-s/.

172 **The child will carry mostly:** Ian Sample, "UK Doctors Select First
Women to Have 'Three-Person Babies,'" *Guardian* (US edition) online, last modified February 1, 2018, https://www.theguardian.com
/science/2018/feb/01/permission-given-to-create-britains-first
-three-person-babies.

172 **Excessive contacts:** J. Valades et al, "ER Lipid Defects in Neuro-

peptidergic Neurons Impair Sleep Patterns in Parkinson's Diseases," *Neuron* 98, no. 6 (June 27, 2018): 1155–69, https://doi.org/10.1016/j.neuron.2018.05.022.

173 **Perhaps no other structure:** N. Sun, R. J. Youle, and T. Finkel, "The Mitochondrial Basis of Aging," *Molecular Cell* 61, no. 5 (March 3, 2016): 654–66, https://doi.org/10.1016/j.molcel.2016.01.028.

173 **In 1954:** D. Harman, "Origin and Evolution of the Free Radical Theory of Aging: A Brief Personal History, 1954–2009," *Biogerontology* 10, no. 6 (December 2009): 773–81, https://doi.org/10.1007/s10522-009-9234-2.

174 **Harman's idea:** R. S. Sohal and R. Weindruch, "Oxidative Stress, Caloric Restriction, and Aging," *Science* 273, no. 5271 (July 5, 1996): 59–63, https://doi.org/10.1126/science.273.5271.59.

174 **Over time, they damage:** E. R. Stadtman, "Protein Oxidation and Aging," *Free Radical Research* 40, no. 12 (December, 2006): 1250–58, https://doi.org/10.1080/10715760600918142.

175 **Strains of mice that made:** S. E. Schriner et al., "Extension of Murine Life Span by Overexpression of Catalase Targeted to Mitochondria," *Science* 308, no. 5730 (June 24, 2005): 1909–11, https://doi.org/10.1126/science.1106653.

175 **As recently as 2022:** J. Hartke et al., "What Doesn't Kill You Makes You Live Longer—Longevity of a Social Host Linked to Parasite Proteins," *bioRxiv* (2022): https://doi.org/10.1101/2022.12.23.521666.

175 **One way they may minimize:** A. Rodríguez-Nuevo et al., "Oocytes Maintain ROS-free Mitochondrial Metabolism by Suppressing Complex I," *Nature* 607, no. 7920 (July 2022): 756–61, https://doi.org/10.1038/s41586-022-04979-5.

175 **Alas, although there were isolated reports:** G. Bjelakovic et al., "Mortality in Randomized Trials of Antioxidant Supplements for Primary and Secondary Prevention: Systematic Review and Meta-analysis," *Journal of the American Medical Association* (*JAMA*) 297, no. 8 (2007): (February 28, 2007): 842–57, https://doi.org/10.1001/jama.297.8.842.

176 **But over the last ten to fifteen years:** S. Hekimi, J. Lapointe, and Y. Wen, "Taking a 'Good' Look at Free Radicals in the Aging Process," *Trends in Cell Biology* 21, no. 10 (October 2011): 569–76, https://doi.org/10.1016/j.tcb.2011.06.008. There are also first-rate discussions of the evidence in López-Otín et al., "Hallmarks of Aging," 1194–217, and A. Bratic and N. G. Larsson, "The Role of Mitochondria in Aging," *Journal of Clinical Investigation* 123, no. 3 (March 2013): 951–57, https://doi.org/10.1172/JCI64125.

176 **Studies with other animals:** See the papers cited in Bratic and Larsson, "Role of Mitochondria," 951–57.

176 **In fact, contrary to the report:** V. I. Pérez et al., "The Overexpression of Major Antioxidant Enzymes Does Not Extend the Lifespan of Mice," *Aging Cell* 8, no. 1 (February 2009): 73–75, https://doi.org/10.1111/j.1474-9726.2008.00449.x.

176 **Giving them a herbicide:** W. Yang and S. Hekimi, "A Mitochondrial Superoxide Signal Triggers Increased Longevity in *Caenorhabditis elegans*," *PLoS Biology* 8, no. 12 (December 2010): e1000556, https://doi.org/10.1371/journal.pbio.1000556.

176 **The naked mole rat lives:** B. Andziak et al., "High Oxidative Damage Levels in the Longest-Living Rodent, the Naked Mole-Rat," *Aging Cell* 5, no. 6 (December 2006): 463–71, https://doi.org/10.1111/j.1474-9726.2006.00237.x; F. Saldmann et al., "The Naked Mole Rat: A Unique Example of Positive Oxidative Stress," *Oxidative Medicine and Cellular Longevity* 2019 (February 7, 2019): 4502819, https://doi.org/10.1155/2019/450281.9.

176 **This may be an example of something called hormesis:** V. Calabrese et al., "Hormesis, Cellular Stress Response and Vitagenes as Critical Determinants in Aging and Longevity," *Molecular Aspects of Medicine* 32, nos. 4–6 (August–December 2011): 279–304, https://doi.org/10.1016/j.mam.2011.10.007.

177 **At the age of about sixty weeks:** A. Trifunovic et al., "Premature Ageing in Mice Expressing Defective Mitochondrial DNA Polymerase," *Nature* 429, no. 6990 (May 27, 2004): 417–23, https://doi.org/10.1038/nature02517. This and several other papers published the following year are reviewed in L. A. Loeb, D. C. Wallace, and G. M. Martin, "The Mitochondrial Theory of Aging and Its Relationship to Reactive Oxygen Species Damage and Somatic MtDNA Mutations," *Proceedings of the National Academy of Sciences (PNAS) of the United States of America* 102, no. 52 (December 19, 2005): 18769–70, https://doi.org/10.1073/pnas.0509776102.

177 **There are reports of a complicated interplay:** E. F. Fang et al., "Nuclear DNA Damage Signalling to Mitochondria in Ageing," *Nature Reviews Molecular Cell Biology* 17, no. 5 (May 2016): 308–21, https://doi.org/10.1038/nrm.2016.14; R. H. Hämäläinen et al., "Defects in mtDNA Replication Challenge Nuclear Genome Stability

Through Nucleotide Depletion and Provide a Unifying Mechanism for Mouse Progerias," *Nature Metabolism* 1, no. 10 (October 2019): 958–65, https://doi.org/10.1038/s42255-019-0120-1.

178 **In these cases, clones:** T. E. S. Kauppila, J. H. K. Kauppila, and N. G. Larsson, "Mammalian Mitochondria and Aging: An Update," *Cell Metabolism* 25, no. 1 (January 10, 2017): 57–71, https://doi.org/10.1016/j.cmet.2016.09.017.

178 **The effect is most pronounced:** N. Sun, R. J. Youle, and T. Finkel, "The Mitochondrial Basis of Aging," *Molecular Cell* 61, no. 5 (March 3, 2016): 654–66, https://doi.org/10.1016/j.molcel.2016.01.028.

178 **One characteristic of aging:** C. Franceschi et al., "Inflamm-aging. An Evolutionary Perspective on Immunosenescence," *Annals of the New York Academy of Sciences* 908, no. 1 (June 2000): 244–54, https://doi.org/10.1111/j.1749-6632.2000.tb06651.x.

179 **Some proteins can sense:** N. P. Kandul et al., "Selective Removal of Deletion-Bearing Mitochondrial DNA in Heteroplasmic Drosophila," *Nature Communications* 7 (November 14, 2016): art. 13100, https://doi.org/10.1038/ncomms13100.

179 **The inhibition of TOR:** M. Morita et al., "mTORC1 Controls Mitochondrial Activity and Biogenesis Through 4E-BP-Dependent Translational Regulation," *Cell Metabolism* 18, no. 5 (November 5, 2013): 698–711, https://doi.org/10.1016/j.cmet.2013.10.001.

179 **In studies, the increased mitochondrial activity:** B. M. Zid et al., "4E-BP Extends Lifespan upon Dietary Restriction by Enhancing Mitochondrial Activity in *Drosophila*," *Cell* 139, no. 1 (October 2, 2009): 149–60, https://doi.org/10.1016/j.cell.2009.07.034.

179 **Besides TOR, other signals:** C. Cantó and J. Auwerx, "PGC-1α, SIRT1 and AMPK, an Energy Sensing Network That Controls Energy Expenditure," *Current Opinion in Lipidology* 20, no. 2 (April 2009): 98–105, https://doi.org/10.1097/mol.0b013e328328d0a4.

179 **Sometimes, though, this effort is futile:** C. Cantó and J. Auwerx, "PGC-1α, SIRT1 and AMPK, an Energy Sensing Network That Controls Energy Expenditure," *Current Opinion in Lipidology* 20, no. 2 (April 2009): 98–105, https://doi.org/10.1097/mol.0b013e328328d0a4.

179 **Physical activity turns on:** See Sun, Youle, and Finkel, "Mitochondrial Basis of Aging," 654–66; J. L. Steiner et al., "Exercise Training Increases Mitochondrial Biogenesis in the Brain," *Journal of Applied Physiology* 111, no. 4 (October 2011): 1066–71, https://doi.org/10.1152/japplphysiol.00343.2011.

180 **One way it spurs mitochondrial function:** Z. Radak, H. Y. Chung, and S. Goto, "Exercise and Hormesis: Oxidative Stress-Related Adaptation for Successful Aging," *Biogerontology* 6, no. 1 (2005): 71–75, https://doi.org/10.1007/s10522-004-7386-7.

180 **Of course, exercise does far more:** G. C. Rowe, A. Safdar, and Z. Arany, "Running Forward: New Frontiers in Endurance Exercise Biology," *Circulation* 129, no. 7 (February 18, 2014): 798–810, https://doi.org/10.1161/circulationaha.113.001590.

180 **But it is better repaired:** J. B. Stewart and N. G. Larsson, "Keeping mtDNA in Shape Between Generations," *PLoS Genetics* 10, no. 10 (October 9, 2014): e1004670, https://doi.org/10.1371/journal.pgen.1004670.

180 **Nevertheless, selection is not perfect:** Y. Bentov et al., "The Contribution of Mitochondrial Function to Reproductive Aging," *Journal of Assistive Reproduction and Genetics* 28, no. 9 (September 2011): 773–83, https://doi.org/10.1007/s10815-011-9588-7.

10. Aches, Pains, and Vampire Blood

183 **These tumor suppressor genes:** M. Serrano et al., "Oncogenic *ras* Provokes Premature Cell Senescence Associated with Accumulation of p53 and p16^{INK4a}," *Cell* 88, no. 5 (March 7, 1997): 593–602, https://doi.org/10.1016/s0092-8674(00)81902-9; M. Narita and S. W. Lowe, "Senescence Comes of Age," *Nature Medicine* 11, no. 9 (September 2005): 920–22, https://doi.org/10.1038/nm0905-920.

184 **Senescent cells are often produced:** M. Demaria et al., "An Essential Role for Senescent Cells in Optimal Wound Healing Through Secretion of PDGF-AA," *Developmental Cell* 31, no. 6 (December 22, 2014): 722–33, https://doi.org/10.1016/j.devcel.2014.11.012; M. Serrano, "Senescence Helps Regeneration," *Developmental Cell* 31, no. 6 (December 22, 2014): 671–72, https://doi.org/10.1016/j.devcel.2014.12.007.

184 **As damage to our DNA accumulates:** These reviews offer a comprehensive view of senescent cells' role in aging: J. Campisi and F. d'Adda di Fagagna, "Cellular Senescence: When Bad Things Happen to Good Cells," *Nature Reviews Molecular Cell Biology* 8, no. 9 (September 2007): 729–40, https://doi.org/10.1038/nrm2233; J. M. van Deursen, "The Role of Senescent Cells in Ageing," *Nature* 509, no. 7501 (May 22, 2014): 439–46, https://doi.org/10.1038/nature13193; J. Gil, "Cellular Senescence Causes Ageing," *Nature Reviews Molecular Cell Biology* 20 (July 2019): 388, https://doi.org/10.1038/s41580-019-0128-0.

185 **They also lived:** D. J. Baker et al., "Clearance of p16[Ink4a]-Positive Senescent Cells Delays Ageing-Associated Disorders," *Nature* 479, no. 7372 (November 2, 2011): 232–36, https://doi.org/10.1038/nature10600; D. J. Baker et al., "Naturally Occurring p16(Ink4a)-Positive Cells Shorten Healthy Lifespan," *Nature* 530, no. 7589 (February 11, 2016): 184–89, https://doi.org/10.1038/nature16932; see also the commentary by E. Callaway, "Destroying Worn-out Cells Makes Mice Live Longer," *Nature* (February 3, 2016): https://doi.org/10.1038/nature.2016.19287.

185 **When researchers used an oral cocktail:** M. Xu et al., "Senolytics Improve Physical Function and Increase Lifespan in Old Age," *Nature Medicine* 24, no. 8 (August 2018): 1246–56, https://doi.org/10.1038/s41591-018-0092-9.

185 **But this isn't strictly true:** Donavyn Coffey, "Does the Human Body Replace Itself Every 7 Years?," Live Science, last modified July 22, 2022, https://www.livescience.com/33179-does-human-body-replace-cells-seven-years.html; P. Heinke et al., "Diploid Hepatocytes Drive Physiological Liver Renewal in Adult Humans," *Cell Systems* 13, no. 6 (June 15, 2022): 499–507.e12, https://doi.org/10.1016/j.cels.2022.05.001; K. L. Spalding et al., "Dynamics of Hippocampal Neurogenesis in Adult Humans," *Cell* 153, no. 6 (June 6, 2013): 1219–27, https://doi.org/10.1016/j.cell.2013.05.002; A. Ernst et al., "Neurogenesis in the Striatum of the Adult Human Brain," *Cell* 156, no. 5 (February 27, 2014): 1072–83, https://doi.org/10.1016/j.cell.2014.01.044.

187 **This leads to immune system decline:** For a comprehensive discussion of stem cell depletion, see López-Otín et al., "Hallmarks of Aging," 1194–217, https://doi.org/10.1016/j.cell.2013.05.039.

188 **After six weeks, the mice:** A. Ocampo et al., "In Vivo Amelioration of Age-Associated Hallmarks by Partial Reprogramming," *Cell* 167, no. 7 (December 15, 2016): 1719–33.e12, https://doi.org/10.1016/j.cell.2016.11.052.

188 **Not only did the animals:** K. C. Browder et al., "In Vivo Partial Reprogramming Alters Age-Associated Molecular Changes During Physiological Aging in Mice," *Nature Aging* 2, no. 3 (March 2022): 243–53, https://doi.org/10.1038/s43587-022-00183-2; D. Chondronasiou et al., "Multi-omic Rejuvenation of Naturally Aged Tissues by a Single Cycle of Transient Reprogramming," *Aging Cell* 21, no. 3 (March 2022): e13578, https://doi.org/10.1111/acel.13578; D. Gill et al., "Multi-omic Rejuvenation of Human Cells by Maturation Phase

Transient Reprogramming," *eLife* 11 (April 8, 2022): e71624, https://doi.org/10.7554/eLife.71624.

189 **Their DNA methylation:** Y. Lu et al., "Reprogramming to Recover Youthful Epigenetic Information and Restore Vision," *Nature* 588, no. 7836 (December 2020): 124–29, https://doi.org/10.1038/s41586-020-2975-4; see also the news item K. Servick, "Researchers Restore Lost Sight in Mice, Offering Clues to Reversing Aging," *Science* online, last modified December 2, 2020, https://doi.org/10.1126/science.abf9827.

189 **These effects could be reversed:** J.-H. Yang et al., "Loss of Epigenetic Information as a Cause of Mammalian Aging," *Cell* 186, no. 2 (January 19, 2023), https://doi.org/10.1016/j.cell.2022.12.027.

190 **He not only connected two rats:** R. B. S. Harris, "Contribution Made by Parabiosis to the Understanding of Energy Balance Regulation," *Biochimica et Biophysica Acta (BBA)—Molecular Basis of Disease* 1832, no. 9 (September 2013): 1449–55, https://doi.org/10.1016/j.bbadis.2013.02.021.

191 **"If two rats are not adjusted":** C. M. McCay, F. Pope, and W. Lunsford, "Experimental Prolongation of the Life Span," *Journal of Chronic Diseases* 4, no. 2 (August 1956): 153–58, https://www.sciencedirect.com/science/article/abs/pii/0021968156900157. Quoted in an overview of the field by M. Scudellari, "Ageing Research: Blood to Blood," *Nature* 517, no. 7535 (January 22, 2015): 426–29, https://doi.org/10.1038/517426a.

191 **But for some reason:** Scudellari, "Ageing Research," 426–29.

191 **By the same criteria:** M. J. Conboy, I. M. Conboy, and T. A. Rando, "Heterochronic Parabiosis: Historical Perspective and Methodological Considerations for Studies of Aging and Longevity," *Aging Cell* 12, no. 3 (June 2013): 525–30, https://doi.org/10.1111/acel.12065.

191 **He showed that old blood:** S. A. Villeda et al., "The Ageing Systemic Milieu Negatively Regulates Neurogenesis and Cognitive Function," *Nature* 477, no. 7362 (August 31, 2011): 90–94, https://doi.org/10.1038/nature10357; S. A. Villeda et al., "Young Blood Reverses Age-Related Impairments in Cognitive Function and Synaptic Plasticity in Mice," *Nature Medicine* 20, no. 6 (June 2014): 659–63, https://doi.org/10.1038/nm.3569.

192 **the Conboys and Rando pointed out:** Conboy, Conboy, and Rando, "Heterochronic Parabiosis," 525–30.

192 **that were not joined:** J. Rebo et al, "A Single Heterochronic Blood

Exchange Reveals Rapid Inhibition of Multiple Tissues by Old Blood," *Nature Communications* 7, no. 1 (June 10, 2016): art. 13363, https://doi.org/10.1038/ncomms13363.

192   **Such cautionary views:** Rebecca Robbins, "Young-Blood Transfusions Are on the Menu at Society Gala," *Scientific American* online, last modified March 2, 2018, https://www.scientificamerican.com/article/young-blood-transfusions-are-on-the-menu-at-society-gala/.

192   **Alarmed, the US Food and Drug Administration (FDA):** Scott Gottlieb, "Statement from FDA Commissioner Scott Gottlieb, M.D., and Director of FDA's Center for Biologics Evaluation and Research Peter Marks, M.D., Ph.D., Cautioning Consumers Against Receiving Young Donor Plasma Infusions That Are Promoted as Unproven Treatment for Varying Conditions," U.S. Food and Drug Administration, press release, February 19, 2019, https://www.fda.gov/news-events/press-announcements/statement-fda-commissioner-scott-gottlieb-md-and-director-fdas-center-biologics-evaluation-and-0.

193   **"Our patients really want":** Emily Mullin, "Exclusive: Ambrosia, the Young Blood Transfusion Startup, Is Quietly Back in Business," OneZero, last modified November 8, 2019, https://onezero.medium.com/exclusive-ambrosia-the-young-blood-transfusion-startup-is-quietly-back-in-business-ee2b7494b417.

193   **As for old blood, they zeroed in:** J. M. Castellano et al., "Human Umbilical Cord Plasma Proteins Revitalize Hippocampal Function in Aged Mice," *Nature* 544, no. 7651 (April 27, 2017): 488–92, https://doi.org/10.1038/nature22067; H. Yousef et al., "Aged Blood Impairs Hippocampal Neural Precursor Activity and Activates Microglia Via Brain Endothelial Cell VCAM1," *Nature Medicine* 25, no. 6 (June 2019): 988–1000, https://doi.org/10.1038/s41591-019-0440-4.

193   **In a second study:** F. S. Loffredo et al., "Growth Differentiation Factor 11 Is a Circulating Factor That Reverses Age-Related Cardiac Hypertrophy," *Cell* 153, no. 4 (May 9, 2013): 828–39, https://doi.org/10.1016/j.cell.2013.04.015; M. Sinha et al., "Restoring Systemic GDF11 Levels Reverses Age-Related Dysfunction in Mouse Skeletal Muscle," *Science* 344, no. 6184 (May 9, 2014): 649–52, https://doi.org/10.1126/science.1251152; L. Katsimpardi et al., "Vascular and Neurogenic Rejuvenation of the Aging Mouse Brain by Young Systemic Factors," *Science* 344, no. 6184 (May 9, 2014): 630–34, https://doi.org/10.1126/science.1251141. These findings are described in a very accessible article by Carl Zimmer, "Young Blood May Hold Key to Reversing Aging," *New York Times* online, May 4, 2014, https://

www.nytimes.com/2014/05/05/science/young-blood-may-hold-key
-to-reversing-aging.html.

194 **Clearing those senescent cells:** O. H. Jeon et al., "Systemic Induction of Senescence in Young Mice After Single Heterochronic Blood
Exchange," *Nature Metabolism* 4, no. 8 (August 2022): 995–1006,
https://doi.org/10.1038/s42255-022-00609-6.

194 **It turns out that blood:** A. M. Horowitz et al., "Blood Factors Trans-
fer Beneficial Effects of Exercise on Neurogenesis and Cognition to
the Aged Brain," *Science* 369, no. 6500 (July 10, 2020): 167–73, https://
doi.org/10.1126/science.aaw2622.

194 **Rando and Wyss-Coray:** J. O. Brett et al., "Exercise Rejuvenates
Quiescent Skeletal Muscle Stem Cells in Old Mice Through Resto-
ration of Cyclin D1," *Nature Metabolism* 2, no. 4 (April 2020): 307–17,
https://doi.org/10.1038/s42255-020-0190-0.

194 **Although they both stimulated:** M. T. Buckley et al., "Cell Type–
Specific Aging Clocks to Quantify Aging and Rejuvenation in Regenerative Regions of the Brain," *Nature Aging* 3 (January 2023): 121–37,
https://www.nature.com/articles/s43587-022-00335-4.

195 **He went to Resurgence Wellness, a Texas outfit:** David Averre and
Neirin Gray Desai, "Tech Billionaire, 45, Who Spends $2 Million a
Year Trying to Reverse His Ageing Reveals Latest Gadget He Uses
That Puts His Body Through the Equivalent of 20,000 Sit Ups in
30 Minutes," *Daily Mail* (London) online, last modified April 5, 2023,
https://www.dailymail.co.uk/news/article-11942581/Tech-billionaire
-45-spends-2million-year-trying-reverse-ageing-reveals-latest-gadget
.html; Orianna Rosa Royle, "Tech Billionaire Who Spends $2 Million a
Year to Look Young Is Now Swapping Blood with His 17-Year-Old Son
and 70-Year-Old Father," *Fortune* online, last modified May 23, 2023,
https://fortune.com/2023/05/23/bryan-johnson-tech-ceo-spends-2
-million-year-young-swapping-blood-17-year-old-son-talmage-70
-father/; Alexa Mikhail, "Tech CEO Bryan Johnson admits he saw 'no
benefits' after controversially injecting his son's plasma into his body
to reverse his biological age," *Fortune*, July 8, 2023, https://fortune.com
/well/2023/07/08/bryan-johnson-plasma-exchange-results-anti-aging/.

11. Crackpots or Prophets?

197 **An entire field of biology:** S. Bojic et al., "Winter Is Coming: The
Future of Cryopreservation," *BMC Biology* 19, no. 1 (March 24, 2021):
56, https://doi.org/10.1186/s12915-021-00976-8.

197 **The idea has been around a long time:** Paul Vitello, "Robert C. W. Ettinger, a Proponent of Life After (Deep-Frozen) Death, Is Dead at 92," *New York Times* online, July 29, 2011, https://www.nytimes .com/2011/07/30/us/30ettinger.html; Associated Press, "Cryonics Pioneer Robert Ettinger Dies," *Guardian* (US edition) online, last modified July 26, 2011, https://www.theguardian.com/science/2011 /jul/26/cryonics-pioneer-robert-ettinger-dies.

199 **One such proponent is Elon Musk:** See "Elon Musk on Cryonics," Elon Musk, interviewed by Zach Latta, YouTube video, 2:09, up-loaded by Hack Club on May 4, 2020, https://www.youtube.com /watch?v=MSIjNKssXAc.

199 **"die on Mars":** Daniel Terdiman, "Elon Musk at SXSW: 'I'd Like to Die on Mars, Just Not on Impact,'" CNET, last modified March 9, 2013, https://www.cnet.com/culture/elon-musk-at-sxsw-id-like-to-die -on-mars-just-not-on-impact/.

200 **It would be like trying to deduce the entire state of a country:** See a particularly cutting article that deals with this and the general issue of cryonics by the neurobiologist Michael Hendrick, "The False Science of Cryonics," *MIT Technology Review*, September 15, 2015, https://www.technologyreview.com/2015/09/15/109906/the-false -science-of-cryonics.

201 **What would be the point:** Albert Cardona, conversation with the author, January 12, 2023.

201 **She took the matter to court:** Owen Bowcott and Amelia Hill, "14-Year-Old Girl Who Died of Cancer Wins Right to Be Cryogen-ically Frozen," *Guardian* (US edition) online, last modified Novem-ber 18, 2016, https://www.theguardian.com/science/2016/nov/18 /teenage-girls-wish-for-preservation-after-death-agreed-to-by-court.

201 **This elicited an outcry:** Alexandra Topping and Hannah Devlin, "Top UK Scientist Calls for Restrictions on Marketing Cryonics," *Guardian* (US edition) online, last modified November 18, 2016, https://www .theguardian.com/science/2016/nov/18/top-uk-scientist-calls-for -restrictions-on-marketing-cryonics.

201 **In almost a mirror image:** Tom Verducci, "What Really Happened to Ted Williams?," *Sports Illustrated* online, last modified August 18, 2003, https://vault.si.com/vault/2003/08/18/what-really-happened-to -ted-williams-a-year-after-the-jarring-news-that-the-splendid-splinter -was-being-frozen-in-a-cryonics-lab-new-details-including-a-decapita tion-suggest-that-one-of-americas-greatest-heroes-may-never-rest-in.

202 **According to press reports:** See sources cited in https://en.wiki

pedia.org/wiki/List_of_people_who_arranged_for_cryonics; when I wrote to Nick Bostrom, he replied, "It has been thus reported in the media. My general stance however has been not to comment on my funereal or other posthumous arrangements . . .", email January 11, 2023.

202 **a San Francisco company called Nectome:** Antonio Regalado, "A Startup Is Pitching a Mind-Uploading Service That Is '100 Percent Fatal,'" *MIT Technology Review* online, last modified March 13, 2018, https://www.technologyreview.com/2018/03/13/144721/a-startup -is-pitching-a-mind-uploading-service-that-is-100-percent-fatal/.

203 **In response, Robert McIntyre, the founder of Nectome said:** Sharon Begley, "After Ghoulish Allegations, a Brain-Preservation Company Seeks Redemption," *Stat* (online), January 30, 2019, https:// www.statnews.com/2019/01/30/nectome-brain-preservation -redemption.

203 **He began his career:** Evelyn Lamb, "Decades-Old Graph Problem Yields to Amateur Mathematician," Quanta, last modified April 17, 2018, https://www.quantamagazine.org/decades-old-graph-problem -yields-to-amateur-mathematician-20180417/.

203 **He asserts that the first humans:** Aubrey de Grey, "A Roadmap to End Aging," TED Talk, July 2005, 22:35, https://www.ted.com/talks /aubrey_de_grey_a_roadmap_to_end_aging/.

203 **if we crack seven key problems:** A. D. de Grey et al., "Time to Talk SENS: Critiquing the Immutability of Human Aging," *Annals of the New York Academy of Sciences* 959, no. 1 (April 2002): 452–62, discussion 463, https://doi.org/10.1111/j.1749–6632.2002.tb02115.x; A. D. de Grey, "The Foreseeability of Real Anti-Aging Medicine: Focusing the Debate," *Experimental Gerontology* 38, no. 9 (September 1, 2013): 927–34, https://doi.org/10.1016/s0531-5565(03)00155-4.

204 **In response to his claims:** H. Warner et al., "Science Fact and the SENS Agenda: What Can We Reasonably Expect from Ageing Research," *EMBO Reports* 6, no. 11 (November 2005): 1006–8, https:// doi.org/10.1038/sj.embor.7400555.

204 **Other mainstream researchers:** Estep et al., "Life Extension Pseudoscience and the SENS Plan," *MIT Technology Review*, 2006, http:// www2.technologyreview.com/sens/docs/estepetal.pdf; Sherwin Nuland, "Do You Want to Live Forever?," *MIT Technology Review* online, last modified February 1, 2005, https://www.technologyreview .com/2005/02/01/231686/do-you-want-to-live-forever/.

204 **One of them, Richard Miller:** Richard Miller, open letter to Aubrey de Grey, *MIT Technology Review* online, November 29, 2005, https://www.technologyreview.com/2005/11/29/274243/debating -immortality/.

205 **"There's going to be much less difference":** Comments by Aubrey de Grey in *The Immortalists*, ibid.

205 **He denied the allegations:** Analee Armstrong, "Anti-Aging Foundation SENS Fires de Grey After Allegations He Interfered with Investigation into His Conduct," Fierce Biotech, last modified August 23, 2021, https://www.fiercebiotech.com/biotech/anti-aging-foundation -sens-turfs-de-grey-after-allegations-he-interfered-investigation -into.

205 **A company report:** SENS Research Foundation, "Announcement from the SRF Board of Directors," news release, March 23, 2022, https:// www.sens.org/announcement-from-the-srf-board-of-directors/.

205 **De Grey, undaunted:** "Meet the Team," LEV Foundation online, accessed August 7, 2023, https://www.levf.org/team.

205 **For example, he has predicted:** David Sinclair, quoted in Antonio Regalado, "How Scientists Want to Make You Young Again," *MIT Technology Review* online, last modified October 25, 2022, https://www .technologyreview.com/2022/10/25/1061644/how-to-be-young -again/.

205 **Such statements:** Catherine Elton, "Has Harvard's David Sinclair Found the Fountain of Youth," *Boston* online, last modified October 29, 2019, https://www.bostonmagazine.com/health/2019/10/29/david -sinclair/.

206 **I doubt whether:** David Sinclair and Matthew LaPlante, *Lifespan: Why We Age, and Why We Don't Have To* (New York: Atria Books, 2019). For a sharply critical review of the book, see C. A. Brenner, "A Science-Based Review of the World's Best-Selling Book on Aging," *Archives of Gerontology and Geriatrics* 104 (January 2023): art. 104825, https://doi.org/10.1016/j.archger.2022.104825.

206 **In an essay on LinkedIn:** David Sinclair, "This Is Not an Advice Article," LinkedIn, last modified June 25, 2018, https://www.linkedin .com/pulse/advice-article-david-sinclair.

207 **They would often make:** As one of hundreds of examples, see this description of companies founded in response to findings on blood transfusions: Rebecca Robbins, "Young-Blood Transfusions Are on the Menu at Society Gala," *Scientific American* online, last modified March 2, 2018, https://www.scientificamerican.com/article/young -blood-transfusions-are-on-the-menu-at-society-gala/.

207 **Even back in 2002**: S. J. Olshansky, L. Hayflick, and B. A. Carnes, "Position Statement on Human Aging," *Journals of Gerontology: Series A* 57, no. 8 (August 1, 2002): B292–97, https://doi.org/10.1093 /gerona/57.8.b292. A total of fifty-one gerontologists cosigned the statement, and the three lead authors also published a popular summary, "Essay: No Truth to the Fountain of Youth," *Scientific American* 286, no. 6 (June 2002): 92–95, https://doi.org/10.1038/scientific american0602-92.

209 **California tech billionaires, especially:** See, for example, Todd Friend, "Silicon Valley's Quest to Live Forever," *New Yorker* online, last modified March 27, 2017, https://www.newyorker.com/mag azine/2017/04/03/silicon-valleys-quest-to-live-forever; Anjana Ahuja, "Silicon Valley's Billionaires Want to Hack the Ageing Process," *Financial Times* online, last modified September 7, 2021, https://www .ft.com/content/24849908-ac4a-4a7d-b53c-847963ac1228; Anjana Ahuja, "Can We Defeat Death?," *Financial Times* online, last modified October 29, 2021, https://www.ft.com/content/60d9271c-ae0a-4d44 -8b11-956cd2e484a9.

210 **When they were young, they wanted to be rich:** This paraphrases an idea expressed previously by Antonio Regalado, "Meet Altos Labs, Silicon Valley's Latest Wild Bet on Living Forever," *MIT Technology Review* online, last modified September 4, 2021, https://www.tech nologyreview.com/2021/09/04/1034364/altos-labs-silicon-valleys -jeff-bezos-milner-bet-living-forever/.

210 **Recently, he wrote a tract:** Yuri Milner, *Eureka Manifesto*, available for downloading at https://yurimilnermanifesto.org/.

211 **When news of Altos Labs:** Antonia Regalado, "Meet Altos Labs, Silicon Valley's Latest Wild Bet on Living Forever," *MIT Technology Review* online, last modified September 4, 2021, https://www.tech nologyreview.com/2021/09/04/1034364/altos-labs-silicon-valleys -jeff-bezos-milner-bet-living-forever/.

211 **Rick Klausner, its chief scientist:** Hannah Kuchler, "Altos Labs Insists Mission Is to Improve Lives Not Cheat Death," *Financial Times* online, last modified January 23, 2022, https://www.ft.com/content /f3bceaf2-0d2f-4ec7-b767-693bf01f9630.

211 **"Our goal is for everyone":** The author was present at the launch of the Cambridge campus of Altos Labs on June 22, 2022.

212 **"I went through a period":** Michael Hall, email message to the author, September 2, 2021.

213 **Other drugs aim to target:** A more comprehensive list of strategies and drugs that are used to combat aging is found in Partridge, Fuentealba, and Kennedy, "Quest to Slow Ageing," 513–32.

213 **Some of the biggest excitement:** M. Eisenstein, "Rejuvenation by Controlled Reprogramming Is the Latest Gambit in Anti-Aging," *Nature Biotechnology* 40, no. 2 (February 2022): 144–46, https://doi .org/10.1038/d41587-022-00002-4.

214 **"Despite intensive study":** Olshansky, Hayflick, and Carnes, "Position Statement," B292–97.

214 **In addition to epigenetic changes:** K. S. Kudryashova et al., "Aging Biomarkers: From Functional Tests to Multi-Omics Approaches," *Proteomics* 20, nos. 5/6 (March 2020): art. E1900408, https://doi .org/10.1002/pmic.201900408; Buckley et al., "Cell Type–Specific Aging Clocks."

216 **This goal was termed:** Kudryashova et al., "Aging Biomarkers: From Functional Tests to Multi-Omics Approaches"; Buckley et al., "Cell Type–Specific Aging Clocks."

217 **"forever remain quixotic":** A. D. de Grey, "The Foreseeability of Real Anti-Aging Medicine: Focusing the Debate," *Experimental Gerontology* 38, no. 9 (September 1, 2003): 927–34, https://doi.org/10.1016 /s0531-5565(03)00155-4.

217 **If anything, data:** "Health State Life Expectancies, UK: 2018 to 2020," Office of National Statistics (UK) online, last modified March 4, 2022, https://www.ons.gov.uk/peoplepopulationandcommunity /healthandsocialcare/healthandlifeexpectancies/bulletins/health statelifeexpectanciesuk/latest.

217 **A United Nations report:** Jean-Marie Robine, "Aging Populations: We Are Living Longer Lives, But Are We Healthier?," United Nations Department of Economic and Social Affairs, Population Division, online, September 2021, https://desapublications.un.org/file/653/download.

218 **A farmer was merrily riding:** Oliver Wendell Holmes, *The Deacon's Masterpiece or the Wonderful One-Hoss Shay*, Cambridge, MA: Houghton, Mifflin, 1891. With illustrations by Howard Pyle. Reproduced in http://www.ibiblio.org/eldritch/owh/shay.html.

219 **Thomas Perls:** Perls, email, November 27, 2021.

219 **This would argue in favor:** S. L. Andersen et al., "Health Span Approximates Life Span Among Many Supercentenarians: Compression of Morbidity at the Approximate Limit of Life Span," *Journals of Gerontology: Series A* 67, no. 4 (April 2012): 395–405 (2012), https:// doi.org/10.1093/gerona/glr223.

219 **Similarly, a variant of a gene:** P. Sebastiani et al., "A Serum Protein Signature of APOE Genotypes in Centenarians," *Aging Cell* 18, no. 6 (December 2019): e13023, https://doi.org/10.1111/acel.13023; B. N. Ostendorf et al., "Common Germline Variants of the Human APOE Gene Modulate Melanoma Progression and Survival," *Nature Medicine* 26, no. 7 (July 2020): 1048–53, https://doi.org/10.1038/s41591 -020-0879-3; B. N. Ostendorf et al., "Common Human Genetic Variants of APOE Impact Murine COVID-19 Mortality," *Nature* 611, no. 7935 (November 2022): 346–51, https://doi.org/10.1038 /s41586-022-05344-2.

12. Should We Live Forever?

221 **The share of older people:** United Nations Department of Economic and Social Affairs, Population Division, *World Population Prospects 2022: Summary of Results* (New York: United Nations, 2022), https://www .un.org/development/desa/pd/sites/www.un.org.development.desa.pd /files/wpp2022_summary_of_results.pdf.

222 **In both social and economic terms:** David E. Boom and Leo M. Zucker, "Aging Is the Real Population Bomb," *Finance & Development* online, June 2022, 58–61, https://www.imf.org/en/Publications /fandd/issues/Series/Analytical-Series/aging-is-the-real-population -bomb-bloom-zucker.

223 **The poor not only live:** Veena Raleigh, "What Is Happening to Life Expectancy in England?," King's Fund online, last modified August 10, 2022, https://www.kingsfund.org.uk/publications/whats-happening -life-expectancy-england.

223 **Things are even worse in the United States:** R. Chetty et al., "The Association Between Income and Life Expectancy in the United States, 2001–2014," *Journal of the American Medical Association* (*JAMA*) 315, no. 16 (April 26, 2016): 1750–66, https://doi.org/10.1001/jama .2016.4226.

223 **Advances in medicine:** V. J. Dzau and C. A. Balatbat, "Health and Societal Implications of Medical and Technological Advances," *Science Translational Medicine* 10, no. 463 (October 17, 2018): eaau4778, https://doi.org/10.1126/scitranslmed.aau4778; D. Weiss et al. "Innovative Technologies and Social Inequalities in Health: A Scoping Review of the Literature," *PLoS One* 13, no. 4 (April 3, 2018): e0195447 (2018), https://doi.org/10.1371/journal.pone.0195447; Fiona Mc-

Millan, "Medical Advances Can Exacerbate Inequality," *Cosmos* online, last modified October 21, 2018, https://cosmosmagazine.com/people/medical-advances-can-exacerbate-inequality/.

224 **This is because fertility:** D. R. Gwatkin and S. K. Brandel, "Life Expectancy and Population Growth in the Third World," *Scientific American* 246, no. 5 (May 1982): 57–65, https://doi.org/10.1038/scientificamerican0582-57.

225 **Elon Musk believes:** Tweet by Elon Musk, August 26, 2022, https://twitter.com/elonmusk/status/1563020169160851456.

225 **Nevertheless, as people live longer:** J. R. Goldstein and W. Schlag, "Longer Life and Population Growth," *Population and Development Review* 25, no. 4 (December 1999): 741–47, https://doi.org/10.1111/j.1728-4457.1999.00741.x.

226 **Large percentages of people:** Paul Root Wolpe, quoted in Jenny Kleeman, "Who Wants to Live Forever? Big Tech and the Quest for Eternal Youth," *New Statesman* online, last modified October 13, 2021, https://www.newstatesman.com/long-reads/2022/12/live-forever-big-tech-search-quest-eternal-youth-long-read.

226 **In 2023:** Angelique Chrisafis, "More Than 1.2 Million March in France over Plan to Raise Pension Age to 64," *Guardian* (US edition) online, last modified March 7, 2023, https://www.theguardian.com/world/2023/mar/07/nationwide-strikes-in-france-over-plan-to-raise-pension-age-to-64.

226 **Reacting to the French protests:** Annie Lowrey, "The Problem with the Retirement Age Is That It's Too High," *Atlantic* online, last modified April 15, 2023, https://www.theatlantic.com/ideas/archive/2023/04/social-security-benefits-france-pension-protests/673733/.

228 **However, at a Hay Literary Festival event:** Interview on Channel 4 (UK), May 27, 2005.

229 **Ishiguro posited a theory:** Kazuo Ishiguro, email to the author, August 6, 2021.

230 **Most studies say our general cognitive abilities:** T. A. Salthouse, "When Does Age-Related Cognitive Decline Begin?," *Neurobiology of Aging* 30, no. 4 (April 2009): 507–14, https://doi.org/10.1016/j.neurobiolaging.2008.09.023; L. G. Nilsson et al., "Challenging the Notion of an Early-Onset of Cognitive Decline," *Neurobiology of Aging* 30, no. 4 (April 2009): 521–24, discussion 530, https://doi.org/10.1016/j.neurobiolaging.2008.11.013; T. Hedden and J. D. Gabrieli, "Insights into the Ageing Mind: A View from Cognitive Neuroscience," *Nature*

Reviews Neuroscience 5, no. 2 (February 2004): 87–96, https://doi.org /10.1038/nrn1323.

230 **The one category:** A. Singh-Manoux et al., "Timing of Onset of Cognitive Decline: Results from Whitehall II Prospective Cohort Study," *BMJ* 344, no. 7840 (January 5, 2012): d7622, https://doi.org/10.1136 /bmj.d7622.

230 **The latter declines steadily:** D. Murman, "The Impact of Age on Cognition," *Seminars in Hearing* 36, no. 3 (2015): 111–21, https://doi .org/10.1055/s-0035-1555115.

231 **This is partly because:** Household total wealth in Great Britain: April 2018 to March 2020, Office of National Statistics, January 7, 2022, https://www.ons.gov.uk/peoplepopulationandcommunity/per sonalandhouseholdfinances/incomeandwealth/bulletins/totalwealth ingreatbritain/april2018tomarch2020; Donald Hays and Briana Sullivan, The Wealth of Households:2020, United States Census Bureau, August 2022, https://www.census.gov/content/dam/Census/library /publications/2022/demo/p70br-181.pdf.

231 **By contrast, the vast majority:** D. Murman, "The Impact of Age on Cognition," *Seminars in Hearing* 36, no. 3 (2015): 111–21, https://doi .org/10.1055/s-0035-1555115.

232 **"at the peak of their careers":** "Tom Williams, "Oxford Professors 'Forced to Retire' Win Tribunal Case," *Times Higher Education*, March 17, 2023, https://www.timeshighereducation.com/news /oxford-professors-forced-retire-win-tribunal-case.

232 **"I had been telling":** P. B. Moore, "Neutrons, Magnets, and Photons: A Career in Structural Biology," *Journal of Biological Chemistry* 287, no. 2 (January 2012): 805–18, https://doi.org/10.1074/jbc.X111.324509.

233 **The other concluded:** V. Skirbekk, "Age and Individual Productivity: A Literature Survey" (MPIDR working paper WP 2003–028, Max Planck Institute for Demographic Research, Rostock, Ger., August 2003), https://www.demogr.mpg.de/papers/working/wp-2003-028.pdf; C. A. Viviani. et al. "Productivity in Older Versus Younger Workers: A Systematic Literature Review," *Work* 68, no. 3 (2021): 577–618, https://doi.org/10.3233/WOR-203396.o.

234 **There is a lot of evidence:** P. A. Boyle et al., "Effect of a Purpose in Life on Risk of Incident Alzheimer Disease and Mild Cognitive Impairment in Community-Dwelling Older Persons," *Archives of General Psychiatry* 67, no. 3 (March 2010): 304–10, https://doi.org/10.1001 /archgenpsychiatry.2009.208; R. Cohen, C. Bavishi, and A. Rozanski,

"Purpose in Life and Its Relationship to All-Cause Mortality and Cardiovascular Events: A Meta-Analysis," *Psychosomatic Medicine* 78, no. 2 (February/March 2016): 122–33, https://doi.org/10.1097 /PSY.0000000000000274.

234 **Social isolation and loneliness:** A. Steptoe et al., "Social Isolation, Loneliness, and All-Cause Mortality in Older Men and Women," *Proceedings of the National Academy of Sciences (PNAS) of the United States of America* 110, no. 15 (March 25, 2013): 5797–801, https://doi .org/10.1073/pnas.1219686110; J. Holt-Lunstad et al., "Loneliness and Social Isolation as Risk Factors for Mortality: A Meta-Analytic Review," *Perspectives on Psychological Science* 10, no. 2 (March 2015): 227–37, https://doi.org/10.1177/1745691614568352.

236 **Arieff believes:** Allison Arieff, "Life Is Short. That's the Point," *New York Times* online, August 18, 2018, https://www.nytimes .com/2018/08/18/opinion/life-is-short-thats-the-point.html.

238 **The clear-eyed view:** *Report: Living to 120 and Beyond: Americans' Views on Aging, Medical Advances and Radical Life Extension* (Washington, DC: Pew Research Center, August 6, 2013), https://www .pewresearch.org/religion/2013/08/06/living-to-120-and-beyond -americans-views-on-aging-medical-advances-and-radical-life -extension/.

Index

NOTE: *Italic page numbers* indicate photographs.